BIRDS - EVOLUTION, BEHA

FORESTED HABITATS AND HUMAN-MODIFIED LAND-USE EFFECTS ON AVIAN DIVERSITY

Birds - Evolution, Behavior and Ecology

Trends in Ornithology Research
Pedro K. Ulrich and Julien H. Willett (Editors)
2010. ISBN: 978-1-60876-454-9

The Role of Nepotism and Competition for the Evolution of Avian Families
Michael Griesser and Jonathan Barnaby
2010. ISBN: 978-1-60876-867-7

Forested Habitats and Human-Modified Land-Use Effects on Avian Diversity
Alvaro Redondo-Brenes and Florencia Montagnini
2010. ISBN: 978-1-60876-879-0

BIRDS - EVOLUTION, BEHAVIOR AND ECOLOGY

FORESTED HABITATS AND HUMAN-MODIFIED LAND-USE EFFECTS ON AVIAN DIVERSITY

ALVARO REDONDO-BRENES

AND

FLORENCIA MONTAGNINI

EDITOR

Nova Science Publishers, Inc.

New York

For permission to use material from this book please contact us:
Telephone 631-231-7269; Fax 631-231-8175
Web Site: http://www.novapublishers.com

Library of Congress Cataloging-in-Publication Data

Forested habitats and human-modified land-use effects on avian diversity / Alvaro Redondo-Brenes and Florencia Montagnini.
p. cm.
Includes bibliographical references and index.
ISBN 978-1-60876-879-0 (softcover)
1. Forest birds--Effect of forest management on--Costa Rica. 2. Birds--Habitat--Conservation--Costa Rica. 3. Rain forest conservation--Costa Rica. 4. Land use--Environmental aspects--Costa Rica. 5. Bird surveys--Costa Rica. I. Montagnini, Florencia, 1950- II. Title.
QL687.C8R43 2009
598.173097286--dc22
2009050562

Published by Nova Science Publishers, Inc. † New York

CONTENTS

PREFACE

As a result of the limitations to protected areas that provide habitat for many wildlife species, in the last two decades efforts have shifted to studying wildlife conservation in human-dominated landscapes. The present conservation study was done in the Path of the Tapir Biological Corridor (PTBC), Costa Rica. The corridor, an 82,000 ha area of fragmented forests, encompasses 55 rural communities with more than 10,000 people. Deforestation and development are the main biodiversity threats in the region. The main objective of this study was to estimate the contribution of ten habitat types: forested areas (wildlife refuges and biological reserves), agroforestry systems (homegardens and agrosilvopastoral systems), and other human-dominated land uses on the conservation of bird species in the corridor. Bird data was obtained using point counts along a whole year. Every habitat type had a total of 20 sample points (200 points total), and each sampling point was visited three times during both the summer and the winter seasons. The bird species were categorized by habitat preference, habitat and feeding guilds, and endangered status. We identified a total of 21,015 bird individuals that corresponded to 293 species. Forest fallows (155 species), forest edges (154), homegardens (148), and agrosilvopastoral systems (143) were where the most species were found on that particular type of land-use. On the other hand, villages (104 species), oil palm plantations (107), residential tourism projects (111), and tree plantations (115) had the lowest bird species 'richness values.' Biological reserves (124 species) and wildlife refuges (121) had intermediate species 'richness values.' They both provided habitat to 68 forest-dependent species that were not found in any other habitat type, and 19 species (70% of the species with a conservation concern) had a priority habitat for conservation in the region. It is also relevant to highlight the importance of homegardens and agrosilvopastoral systems as habitat for 70% of the avian diversity of the region, including the endangered Scarlet Macaw (*Ara macao*). Biodiversity conservation should always include primary and old-secondary forests (e.g., those forests located in wildlife refuges, biological reserves, or isolated forest fragments); as these are the most important habitat type for wildlife at the landscape level. However, the research results also show that the matrix especially agroforestry systems and plantations are important to ensure wildlife conservation in the biological corridor in the short and long term.

ACKNOWLEDGMENTS

We thank ASANA, Hacienda Baru and Rancho La Merced National Wildlife Refuges, Oro Verde Biological Reserve, La Cusinga Ecolodge, the Firestone Center for Restoration Ecology, and the local individuals who collaborated with data collection. Pedro Porras provided training in bird identification to A. Redondo-Brenes before starting this project. Danny and Janan Duarte, Julian Odio, Noel Urena, and Cristian Valenciano also provided important information about birds of the PTBC. This project was funded by the School of Forestry and Environmental Studies, the Tropical Resources Institute, the Council for Latin American and Iberian Studies, the Center for Field Biology Pilot Grant, and the John F. Enders Fellowships and Research Grants, all of them at Yale University; and by the Evergreen Fellow Grant Program through the NGO "Friends of Osa."

INTRODUCTION

The establishment of protected areas is one means of counteracting some of the threats to biodiversity such as habitat loss and fragmentation [1, 2]. As a result of this conservation concern, in the last three decades there has been an exponential growth in the number of protected areas worldwide, particularly in tropical countries where biodiversity is the greatest [3, 4]. However, most of the world's biodiversity exist outside protected areas [5], often because of the socio-economic and political constraints that limit the amount of land allocated to protected status. Protected areas, therefore, do not necessarily hold the extent of habitat needed for most living species, forcing wildlife to seek habitat and refuge beyond the artificial boundaries of parks [2, 5 - 7]. Costa Rica, though it is one of the most diverse countries in the world [8]; and also has one of the most significant systems of protected areas [9], is not an exception. Fragmentation and isolation of Costa Rican protected areas are jeopardizing biological conservation [6, 10]. Therefore, given the limitations of the protected areas system, there is a pressing need to study biodiversity and improve management not only within the protected areas' boundaries, but also in the fragmented landscape matrices that surround them.

This book provides a case study and a theoretical background to understand the effects of fragmentation and human-dominated landscapes on the conservation of wildlife within the tropics. The study concentrates on the fragmented and human-dominated Path of the Tapir Biological Corridor (PTBC) in Costa Rica. This region is one of the most diverse areas of Central America [11], but whose biodiversity is at risk due to deforestation and development. The main goal of this study was to assess the contribution of ten different habitat types for the conservation of bird species in the region. The ten land-use types described are: biological reserves, wildlife refuges, homegardens, agrosilvopastoral systems, forest edges, forest fallows, tree plantations, oil palm plantations, local villages, and residential tourism projects. Aside from the contribution of habitat types to the overall avian diversity of the region, we also evaluated the potential contribution of these habitat types for species of conservation concern.

Chapter 1

BACKGROUND ON LAND-USE CHANGE

LANDSCAPE FRAGMENTATION AND LAND-USE CHANGE EFFECTS ON WILDLIFE SPECIES

Destruction or alteration of natural habitat by humans, resulting in land-use change, represents one of the most serious threats to biodiversity worldwide [9]; along with climate change, and invasive species [12]. Due to this loss of natural habitat there is a challenge of maintaining and conserving biodiversity in fragmented landscapes now dominated by human land-use [13]. Habitat fragmentation has two main components: (1) a reduction in the area covered by a habitat type or natural habitat generally, in a landscape; and (2) a change in habitat configuration, with the remaining habitat left in smaller and more isolated patches [14]. As a consequence of these land-altering processes, Anderson and Jenkins [15] mention that fragmentation may lead to: (1) elimination or dangerous reduction of populations of large, wide-ranging species, including many top carnivores; (2) unraveling of entire biological communities – as, for example, when the decline of top carnivores in fragmented habitats result in the "release" of smaller predators and herbivores, leading to overpredation or overgrazing that may eventually eliminate species or destabilize communities; (3) destruction or degradation of remaining habitat through the intrusion of edge effects such as altered microclimate or invasive species; and (4) disruption of key ecological processes dependent on increasingly rare animal agents – such as pollination, seed dispersal, predator-prey interactions, and nutrient cycling.

The response of wildlife species to habitat destruction and fragmentation is influenced by the species' home range area, body size, food resources and foraging patterns, nesting and shelter requirements, as well as tolerance to habitat disturbance and sensitivity to altered microclimates [13]. Based on these animal characteristics, the species vulnerable to fragmentation can be organized into six groups [9]:

1) *Wide ranging species:* species like large carnivores and migratory ungulates that roam a large area in the course of their daily or seasonal movements. Also animals of heterogeneous landscapes such as amphibians and turtles are vulnerable to fragmentation because they depend on distinct habitats for different phases of their life cycles.
2) *Non-flying species:* species with poor dispersal abilities may not travel far from where their born, or may be stopped by barriers such as a road or a clear-cut. Examples are carrion beetles in Amazonian forests being fragmented by pasture development [16], and arboreal mammals, peccaries, and many insect bats that are unwilling to enter the matrix [17].
3) *Species with specialized requirements:* Species that have specialized habitat or resource requirements are often vulnerable to extinction, especially when those resources are unpredictable in time or space. For example, the great green macaws (*Ara ambigua*) nest and forage more than 90% of the time in almond trees (*Dipteryx panamensis*), and the decrease of these trees has had a negative impact of the macaws populations in Costa Rica [18].
4) *Large-Patch or Interior Species:* Some species occur only in large patches of forests or other habitats, and are absent from small patches with little or no true interior habitat. For example, in Central Amazonia, many forest understory birds–including army ant followers, solitary species, members of mixed-species flocks, and terrestrial species-strongly avoid forest edges [19].
5) *Species with low fecundity:* A species with low reproductive capacity cannot quickly rebuild its population after severe reduction caused by a number of factors. For example, large mammals as the tapir and jaguar have low fecundity in comparison with small mammals like rodents.
6) *Species vulnerable to human exploitation or persecution:* Some species are actively sought by people for food, furs, medicine, pets, or for other uses. Whereas other species, snakes and large predators may be killed on sight. For example, as a result of hunting such species as the jaguar and peccaries have decreased their populations in Corcovado National Park in Costa Rica in the last two decades. Jaguar populations changed from 150

individuals in 1990 to only 30 individuals in 2004 and peccaries from 2000 in 2000 to only 400 in 2004 [20].

Due to the concerns about the effects of habitat fragmentation and isolation on wildlife conservation, there have been an increasing large number of studies on single species or assemblages in fragmented landscapes throughout the world [21; see also a large list of papers summarized by 13]. However, most of those studies have focused on the fragments themselves ignoring species distribution in surrounding sites such as agricultural or other human-dominated areas [22]. Thus, there is a need to study the importance of the landscape matrix for biological conservation.

Importance of the Landscape Matrix for Wildlife Conservation

Assessing the conservation potential of human-dominated landscapes requires investigation of the activities, movement, and persistence of species not only in remnants of native habitats but also in the full array of countryside habitats [14, 23, 24], known also as the matrix. The matrix can be defined as the most widespread habitat within which other elements are embedded, and it can be either the original habitat type (e.g., a primary forest now surrounded by a matrix of agroforestry) or a modified one (e.g., an agricultural plantation surrounded by a matrix of pristine forest) [15]. In anthropogenically modified landscape, some of the matrix or all of it may be human-modified communities such as agricultural fields of various types, agroforestry systems, clear-cut forests, pastures, tree plantations, fallow land, and towns [7, 25].

As part of the landscape the matrix plays different roles for wildlife conservation:

1) *Matrix regulating the movement of organisms:* the matrix has a significant effect on connectivity in forest landscapes and in determining the quantity of successful movements among habitat patches [7, 26, 27], acting as a selective filter (not an absolute barrier) for the movement of species across the landscape [26]. The type of vegetation cover in the matrix determines the pore size of the matrix movement. For example, tall secondary forests are analogous to a filter with a large pore size that allows more faunal movement because of its structure similar to primary

forest. On the contrary, a matrix maintained in pasture has a fine pore size that may impede much faunal movement [26].

2) *Matrix as a resource:* this is the case when the matrix can provide abundant food resources that can be exploited by some species. In addition, the matrix may also offer access to some resources that are rarely needed by a species, perhaps seasonally, such as forage plants rich in a scarce mineral, a favorable hibernation site, or access to a pollinator [7].
3) *Matrix as secondary habitat:* it is possible that the matrix may serve as secondary habitat for some populations [7, 27]. This situation is called a source-sink axis [28]. The source population is one in which reproduction is adequate to balance mortality and usually to export surplus individuals, as well. Such a population supplies the residents with a secondary habitat. Sink populations are those living in secondary habitat [7]. If sink populations produce enough offspring those may supplement populations in reserves [7, 27].
4) *Matrix as a sink or stopper:* the matrix may function as a dispersal sink when the matrix is large relative to the size of patches, it does not support a resident population of a target species, and when few other habitat patches are available to dispersers [7]. In addition, the matrix may act as a "stopper" when it is a complete barrier to dispersal [7].

Several studies addressing the effect of the landscape matrix on wildlife conservation have been conducted in the tropics [22, 24 - 26, 29, 30]. Even though the authors mentioned earlier clarify that there is no substitute for native forest habitat, they also highlight the importance of human-dominated landscapes (e.g., agroforestry systems, coffee plantations, pastures, tree plantations, secondary forests) for the conservation of butterflies [29], non-flying mammals [25, 26], frogs [26], moths [22], and birds [24, 26, 31, 32]. For instance, in a study of non-flying mammal species in the montain areas of Costa Rica, Daily *et al.* [25] found a total of 26 native species. From these 26 species, 9 (35%) were restricted to forest habitat, 14 (54%) occurred in both forest and agricultural habitats, and 3 (11%) were found only in agricultural habitats, showing that the majority of native mammals use countryside habitats. In the same region, Hughes *et al.* [24] found a total of 144 bird species. For this study it was estimated that 46% of those native to the region were utilizing the matrix in some manner. Moreover, it was predicted that bird richness in the matrix would decline to approximately 40% if tall trees and edge habitats were removed from the landscape. Results of the Biological Dynamics of Forest Fragments Project (BDFFP) in central Brazil also report differences in the behavior of wildlife in forest fragments and in the surrounding matrix [26]. Overall, for ants, small mammals, birds, and frogs it was found that between 8-15% of the primary-forest species were recorded in matrix habitats. In addition, for three vertebrate groups (mammals, frogs, and birds) there were positive and significant correlations between population abundance in the matrix and vulnerability to fragmentation. This suggests that species that avoid the matrix tend to decline or disappear in fragments, while those that tolerate or exploit the matrix often remain stable or increase following landscape fragmentation [26].

ENHANCING WILDLIFE CONSERVATION THROUGH BIOLOGICAL CORRIDORS

Minimizing the effects of isolation by enhancing landscape connectivity is one way to counter the adverse effects of fragmentation [13, 33] in addition to increasing the effective habitat area [33]. Connectivity can be defined as linkages of habitats, communities, and ecological processes at multiple spatial and temporal scales [34]. Connectivity is used to describe how the spatial arrangements and the quality of elements in the landscape affect the movement of organisms among habitat patches [35]. Among other alternatives corridors are a strategy for maintaining connectivity in fragmented landscape [36]. A number of definitions and types of corridors have been proposed over time [7, 13, 15, 35]. For instance, Forman [35] defined corridors as strips of habitat that differ from the adjacent habitat on both sides. A more detailed definition provided by Anderson and Jenkins [15] defined corridors as spaces in which connectivity between species, ecosystems, and ecological processes is maintained or restored at various levels. Among the types of corridors there are four classified by Bennett [13]:

1) natural corridors, such as waterways and streams and their associated riparian vegetations;
2) remnant corridors, such as strips of logged forest within clear-cuts, natural woodlands along roadsides, and natural habitat retained as links between nature reserves;
3) regenerated corridors, such as fencerows and hedges; and
4) planted corridors, such as windbreaks or shelterbreaks and urban greenways.

Regardless the type of corridor, Forman [35], mentions that there are six main functions that corridors provide:

1) Conduit: corridors act as conduit when it provides for movement between habitat patches, but organisms do not reside within it [37].
2) Habitat: corridors perform as habitat when the organisms have enough resources for survivorship, reproduction, and movement [37].
3) Filter: corridors act as filter when there is some level of permeability and some organisms and material pass through the corridor whereas others cannot [37].
4) Barrier: Corridors may represent a barrier to animals when the blockage is complete and none of the animals can cross the corridor [37].
5) Source: Corridor can work as a source when it allows animals to emanate from it [37].
6) Sink: poorly designed corridors may act as population sinks, because the large amount of edge exposes animals to predation from matrix dwellers and competition from the other general species [38, 39, cited by 37].

Although biological corridors can work as a good tool for biological conservation, there is a large controversy about their efficacy [13, 36, 40]. Bennett [13] summarizes the critics in three main points:

1) whether sufficient scientific evidence is available to demonstrate the potential conservation benefits of corridors;
2) whether the potential negative effects of corridors may outweigh any conservation value;
3) whether corridors are a cost-effective option in comparison with other ways of using scarce conservation resources.

Several disadvantages have been reported with regard to the corridors, such as serving as pathways for fire, predators, and pathogens. These aspects can undermine conservation objectives [15, 41], and it is clear that corridors are not the solution to all of the current conservation problems [33]. Many of the potential disadvantages could be avoided or mitigated if the proper design is applied to the specific requirements for the species, habitats, ecosystems, and ecological processes of concern in each case [15]. However, further research has to be done in order to investigate the positive or negative impacts of corridors and their elements (land-use types), especially at the landscape level.

Chapter 2

COSTA RICA AND THE PATH OF THE TAPIR BIOLOGICAL CORRIDOR

BIODIVERSITY IN COSTA RICA

Though Costa Rica corresponds to only 0.03 % of the Earth's landmass, it encompasses more than 5 % of the total species worldwide, being one of the 20 most mega-diverse countries in the world [8]. It contains 10,000 vascular plant species (1,500 of them are different kinds of orchids), 864 bird species, 178 amphibian species, 228 reptile species, 236 mammal species, and 1,013 freshwater and marine fish species [8, 42]. Additionally, a Costa Rica's national biodiversity study concluded that there were more than 500,000 species of organisms within the country and where an outstanding 83 % were unknown to science. In specific cases, such as insects, fungus, bacteria, and viruses, the percentage of unknown species may exceed even 90 % [8].

From the total number of described species, 1.4% (1204 species) corresponds to endemic species to Costa Rica. For the vertebrate groups, amphibians have 36 endemic species (20% of the group), reptiles 36 species (16%), freshwater fish 19 species (14%), mammals 6 species (2.5%), and birds 7 species (0.8%) [8]. A total of 1,606 species, corresponding to 18%, are considered either endangered species or threatened species. The amphibian species are the group with the most threatened species with more than 45% of the total group, followed by reptiles (12%), birds (9%), and mammals (6%), respectively [8].

According to Wainwright [43], several factors may contribute to the high biodiversity of Costa Rica. Besides from being located in the tropics, the country has four mountain ranges and varied weather systems with wind blowing in from both oceans at different times of the year, reacting together in diverse ways to create a dense and varied mosaic of habitats. According to Holdridge [44], life zone system in Costa Rica is divided into 23 different zones and/or transition zones. This system uses gradients of temperature and precipitation relative to evapotranspiration and elevation to form a theoretical framework for plant communities or associations. While the tropical wet forests of the Atlantic and southern Pacific lowlands are wet almost year round and receive more than 5,000 mm of rain annually, the tropical dry forests of the northwest have a severe and prolonged dry season and receive just 1,500 mm per year. And while temperatures in this latter region can exceed 30 ºC, the temperature sometimes drops below freezing on the tropical subalpine rain paramos on the high peaks of the Talamanca mountain range (4,000 meters above sea level). Costa Rica's diversity is also influenced by its location right between North and South America. Located as a bridge, the country harbors a blend of species typical of both regions. For example, for mammals, not including the endemic species, 21 species reach the southern limits of their ranges, and 27 species the northern limits in Costa Rica.

DEFORESTATION PATTERNS AND LAND-USE CHANGE IN COSTA RICA

Costa Rica forest cover has dramatically changed in the last century. Until the early 1900s, Costa Rica was almost entirely covered by rich and diverse tropical forests. However, forest cover decreased where its loss totaled more than 80 % of the original forest from the 1980s [45, 46]. This resulted in part from official policies that placed a priority on demographic growth and agricultural production implemented over the 1950s and 1960s [45, 47, 48]. The deforestation process can be evaluated in three periods: forestry regime before the 1970s, the interventionist forestry regime in the 1970s and 1980s, and the forestry regime of the 1990s [45].

FORESTRY REGIME BEFORE THE 1970S

Over this period, forests were considered worthless on the frontier land, and an individual's reputation for hard work depended on the amount of land one cleared [45]. The principle causes of deforestation were reported to be the demand for land rather than for wood [49]. On the Costa Rican frontier, settlers generally occupied land by squatting [45]. Little clearing occurred outside of the central plateau and the western port until the 1950s and 1960s. Then trade between Costa

Rica and the rest of the world started to rise, exposing the economy to ever-changing international prices in various commodities. Agriculture expansion occurred for cattle in the north, coffee in the center, and, as multinationals expanded, also banana plantations in the Atlantic region. Sugar cane plantations, geared to exports and internal market, also arose [48]. As a result of these issues, it was estimated that for the late 1960s the forest cover was only 49 % of the total area [50].

THE INTERVENTIONIST FORESTRY REGIME IN THE 1970S AND 1980S

In the 1970s and 1980s, Costa Rica was mainly in the negative environmental headlines for having one of the highest deforestation rates worldwide. The deforestation rates averaged consistently two to three times higher than the overall regional average for Latin America [51]. Over this period the forest cover changed from 49 % in 1969 to only 17 % in 1983 [52].

Faced with rapidly disappearing forests, the Costa Rican government responded with a multifaceted approach. First, the system of national parks was created in the 1970s. Second, a complex regulatory framework was established governing forestry on privately owned lands through the Forestry Law of 1969 and its revisions of 1973, 1979, and 1986. Third, financial incentives were provided for reforestation, later for natural forest management, and then for forest preservation [45]. Over that period of time, to cut a tree, landowners had to acquire a permit required by the General Forestry Office (DGF in Spanish) through a management plan elaborated by a forester [45].

FORESTRY REGIME OF THE 1990S

A new Forestry Law was enacted in 1996 in Costa Rica. The new law created mechanisms for stakeholder participation in forestry policy making and for the creation and distribution of subsides for preserving forest cover [45] such as the payments for environmental services. The law prohibits cutting trees or converting use of private lands determined to be essential to biodiversity or watershed protection (Article 2). Similarly, if private land is under forest cover, conversion of use (including conversion to plantation) is prohibited (Article 19). The law simplifies the requirements for forest management plans and places restrictions on the time of the year when trees can be harvested (Articles 20-21). It also creates a new subsidy to be paid to landowners in recognition of the uncompensated environmental services provided by forests (Articles 22-27).

The implementation of this new forestry law, along with the rise in ecotourism in the 1990s, and the fall of international markets such as the cattle in the Pacific coast; the northern region slowed the deforestation rate [45, 48]. In fact, the forest cover of Costa Rica rose from 17 % in 1983 [52] to values close to 40 % later in the 1990s [46]; and for 2005, the forest cover was estimated in more than 50% of the country [53]. This increment is due to the implementation of reforestation programs and the establishment of secondary forests on abandoned agricultural lands throughout the country, in part subsidized by payment for environmental services [45, 46, 48, 54].

STUDY AREA DESCRIPTION: PATH OF THE TAPIR BIOLOGICAL CORRIDOR

The Path of the Tapir Biological Corridor (PTBC) is located in southwestern Costa Rica (Figure 1) and is a part of the multinational Mesoamerican Biological Corridor (MBC). The corridor's main objective is to create a network of sites favorable to fauna and flora between the forests of the Osa Peninsula and Golfo Dulce, including Corcovado National Park. These forests connect with those located in the Los Santos Forest Reserve in the Talamanca Mountain Range. The PTBC area covers approximately 82,000 has, inhabited by 10,000 people in 55 communities. This mosaic of human-dominated land uses includes primary and secondary forests, native and exotic forest plantations, agriculture, agroforestry systems, bamboo, oil palm plantations, ecotourism projects, and others.

The corridor is composed of terrain ranging in altitude from 0 to 1100 meters above sea level [11]. According to the Holdridge System [44], three life zones are represented within it: (1) The Tropical Wet Forest that comprises the entire length of the corridor, especially the lower zones; (2) The Tropical Wet Forest Transition to Premontane, represented by a small fringe at the intermediate altitudes, and (3) The Tropical Rainforest Premontane found at the highest elevation in the corridor. The climate in this region of the country is categorized as Seasonal Tropical Wet with an annual precipitation of 4,239 mm and average temperature of 27 ºC. There is also a high precipitation during the rainy season of as much as 685 mm per month (maximum of 1,715 in October 1988), and a pronounced 4 to 5 months dry season with little rain (less than 100 mm) from January to May [11].

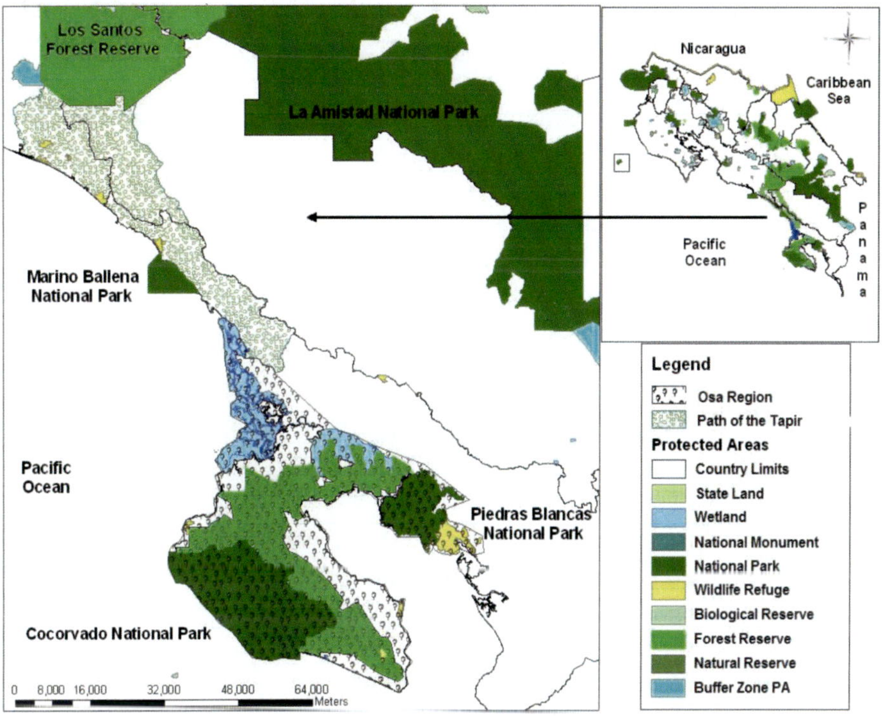

Figure 1. The Path of the Tapir Biological Corridor is located in southwestern Costa Rica and is part of the multinational Mesoamerican Biological Corridor.

The soils in the corridor have a low fertility, and are generally classified as Ultisols and Inceptisols. There are more than 30 rivers among which the main ones are Coronado, Higueron, Uvita, Baru, Hatillo, Savegre, and Guabo [11].

Path of the Tapir Biological Corridor (PTBC) is one of Costa Rica's most diverse regions. A Rapid Ecological Assessment carried out in the region identified a total of 2700 plant species which represent 10% of the total plant diversity in the country, as well as 173 mammal, 26 amphibian, 46 reptile, and 191 bird species [11]. Moreover, 470 avian species have been registered in the area in the six most recent annual bird counts. The PTBC provides habitat to some endangered animal species such as margay (*Leopardus wiedii*), ocelot (*Leopardus pardalis*), jaguar (*Panthera onca*), scarlet macaw (*Ara macao*), spider monkey (*Ateles geoffroyi*), squirrel monkey (*Saimiri oerstedii*), great curassow (*Crax rubra*), and endemic tree species such as quira (*Caryodaphnopsis burgeri*).

DESIGN, PLANNING, AND IMPLEMENTATION OF THE PTBC

Conservation awareness in the PTBC began earlier in the 1980s, and the creation of the local NGO ASANA (Friends of Nature from the Central and South Pacific) for local people was the first step to implement the corridor. ASANA has not only addressed conservation issues, but has also worked to develop community-based sustainable projects. The idea of the PTBC was first under consideration in 1996, but it was not officially recognized until 2000 by Costa Rican authorities. The corridor design encompasses high elevation lands to protect water sources for local consumption and a matrix of high and lowlands to protect habitat for hundreds of species. The PTBC is one of the few projects where a Rapid Ecological Assessment was carried out to justify its importance as a conservation and development tool in Costa Rica.

The strategy for the creation of the PTBC consisted of three elements:

1) Encourage land owners to create private biological reserves such as a) National Wildlife Refuges like Hacienda Barú and La Merced; b) Informal reserves with no particular government recognition such as Oro Verde Biological Reserve and Rafiki Safari Lounge; c) Land held in reserve through payment for environmental services (PES); d) Land protected by environmental easements.
2) Encourage and assist landowners in the protection of their reserves. This is being accomplished through environmental education in surrounding communities, and by creating covenants through which the National Parks Foundation (NPF) hires official park rangers and reserve owners and neighbors reimburse the cost of the salaries to NPF.
3) Encourage the creation of natural corridors along rivers, streams and fences in lands not currently held in reserve, allowing connections between forest patches. This is being accomplished through environmental education and public awareness.

Within their strategy, the PTBC organizations strive to achieve four major goals:

a. Restore natural habitats to enable the return of terrestrial endangered species and marine species such as the leatherback sea turtles (*Dermochelys coriacea*) to the forests and beaches they inhabited half a century ago, and protect existing species by allowing local populations to become more robust;

b. Halt deforestation within the corridor and restore forests in order to complete the corridor and protect water sources for use by communities, particularly in watersheds and groundwater recharge zones;
c. Encourage local communities to value resources and use them for sustainability; and
d. Create interconnecting forest reserves between the existing network of forest and wildlife reserves. Specific conservation programs have been developed by the PTBC organizations to accomplish its goals in the region and they are another important step in corridor implementation [15].

The main programs implemented by PTBC organizations in the last decade are as follows:

1) Environmental education programs in more than 32 schools and workshops for local adults to build a better understanding of sustainable forest management. In addition, within the PTBC there are road signs to caution drivers about the presence of wildlife species and to inform people about the existence of the corridor.
2) Promotion of land conservation areas through Payment for Environmental Services in collaboration with the government.
3) Community-development projects to provide better revenues to local people through sustainable agricultural practices.
4) Coordination to develop drinkable water associations aimed at providing drinkable water and proper watershed management to ensure water to locals.
5) Training and approval of 130 voluntary local game wardens that patrol the regions ad honorem.
6) Monitoring and protection of sea turtles and other flora and fauna species. To date, some of those programs are working well and others still need more funds to be properly implemented. Despite all these conservation programs, there are several threats to biodiversity in the region.

Threats to biodiversity are development and deforestation, illegal logging, poaching and hunting, and lack of law enforcement. This is a very attractive area for foreigners and tourism projects, and as a result, several real estate developers are buying the land from locals, cutting the forest to create new roads, and building residential areas, thus increasing fragmentation. Moreover, even though it is claimed that hunting has decreased in the last three decades, this is still an issue, and hunters are mostly coming from communities outside the corridor. Another main constraint for the PTBC to achieve biological conservation is funding to support more projects. To illustrate, local people in a workshop, held in ASANA's office in June 2006, mentioned that there was a lack of funds to continue some of their current projects or to implement new ones. The threats to biodiversity grow when we append to the described problems a lack of law enforcement, interest to control those problems, funds, and in a few cases corruption from the governmental organizations. In fact, corruption was highlighted by local farmers who were upset for the authorities' negligence to enforce the law against developers and other landowners in the region.

Stakeholder Engagement and Economic Incentives

Large-scale conservation projects in private and human-dominated landscapes require the participation of local people and they have to receive economical incentives to ensure their support [15]. Support to the PTBC project by its participants stems from the fact that they were taken into consideration in the corridor's objectives. For example, since water quality and quantity are important issues in the region [55], and a water conservation approach has proved to be an effective tool for biodiversity conservation as well [56]; it has been an achievement of the corridor leaders to include water as a priority in the region. According to Newcomer [57], the PTBC participants can be grouped into three main in-country clusters: community organizations, government, and landowners. Community organizations include cooperatives, foundations, women's groups, environmental groups, agricultural groups, community development associations, and the PTBC Coordinating Committee. The government is represented by a local liaison between PTBC program and Ministry of Environment (MINAE). National ministries and municipal government agencies have direct interests in and influence on the program [57]. The PTBC consists of three municipalities (Aguirre, Osa, and Perez Zeledon), and three conservation areas (ACOPAC, ACLA-P, and ACOSA) led by MINAE, but to date only the municipality of Aguirre, and the conservation areas ACOPAC, and to a lesser extent ACLA-P, are supporting their conservation and development initiatives. Finally, landowners can be divided into those who participate in conservation and those who don't. The latter, mostly developers, constitute one of the main threats to biodiversity conservation in the area. A fourth group, not mentioned by Newcomer (2002), consists of individuals and institutions funding specific conservation and development projects within the PTBC. Among them are AVINA Foundation,

Cedarena Land Trust, Costa Rica Conservation Trust, the Mesoamerican Biological Corridor Coordinating Committee, and the United Nations Development Programme (UNDP).

The participation of local landowners and their willingness to maintain forest cover will determine the future success of the PTBC project. The fate of the existing biodiversity in the region depends on landowners' management decisions, making attractive incentives necessary to ensure their engagement in conservation practices [15]. The PTBC organizations help landowners find economic incentives for sustainable land management. Some of these are guidance in obtaining PES from the government (to date more than 3,500 ha of forest are under protection through PES), funding for community-development projects, and technical assistance to address water management and other conservation issues in the region. Additionally, local landowners can benefit from conserving their lands through ecotourism.

Despite the work that has been done by the PTBC organizations to match conservation and development in the region, to date, most of the biological data of the corridor is lacking and further research is necessary to assess the conservation of wildlife within this fragmented corridor.

LAND USE IN THE PATH OF THE TAPIR BIOLOGICAL CORRIDOR

Current land use in the corridor has been the result of the process of local and national settlement and development policies. Between 1940 and 1960, Costa Ricans migrated from the northern part of the country toward this region, searching for new lands to clear and cultivate. Between 1960 and 1990, an expanding cattle industry was a major contributing factor to deforestation [11]. To date, deforestation rates have decreased, forest cover has increased as a result of pasture abandonment, and conservation efforts are being undertaken through the establishment of 1 National Park, 5 National Wildlife Refuges, and 27 Informal Private Nature Reserves [Ewing, J. personal communication, June 2006]. The main land uses in the region are primary and secondary forests (35,682 ha), pastures (22,243 ha), young secondary forests (13,488 ha), agriculture (4,074 ha), forest tree plantations (2915 ha), and mangroves (1,160 ha) [11].

Most local people work in agriculture, and recently ecotourism has been growing as another important source of income in the region. However, as land prices have increased, land ownership is shifting from local farmers to wealthy foreigners. Costa Ricans have either become employees of the new owners, or migrated to larger cities in Costa Rica or to the United States and Europe. This poses a new challenge for the region, as protected areas are interspersed with private lands [58]. Therefore, in order to avoid land use change and ensure connectivity among all protected areas, it is essential that the corridor's conservation plan provides adequate incentives to landowners.

Agricultural lands in the region are mostly planted with rice, oil palm, and small scale crops. Pastures for cattle have different characteristics. Some of them have living fences of native tree species, as well as scattered remnant trees as shade for cattle. Tree plantations are mostly of timber species that are being planted in the Caribbean and Pacific lowlands of the country. The main species based on area planted is *Tectona grandis* (teak), an exotic species brought to Costa Rica from Asia. Second, *Terminalia amazonia* (amarillón), a native species that is being planted in medium or small scale plantations. The third species is *Gmelina arborea* (melina) also exotic to Costa Rica. Teak and melina plantations represent more than 50% of the total area of tree plantations in the country [59].

Chapter 3

RESEARCH METHODS

DESCRIPTION OF THE HABITATS USED FOR EVALUATION OF BIRD ABUNDANCE AND DIVERSITY

Based on the predominant land uses currently present in the PTBC we selected ten different habitat types that were located up to 500 meters above sea level. The studied habitat types were wildlife refuges, biological reserves, forest edges, forest fallows, tree plantations, oil palm plantations, homegardens, agrosilvopastoral systems, residential tourism projects, and villages. For each habitat type a total of 20 different sample points were located along the whole corridor, based on 16 routes, using existing roads or trails.

a. *Wildlife Refuges:* We surveyed forested areas (primary or old secondary forest) within two refuges: Rancho La Merced and Hacienda Baru (10 sample points at each refuge for a total of 20 points for this type of habitat). La Merced (329 ha) and Baru (330 ha) are private reserves and have governmental protection since 1995.
b. *Biological reserves:* We selected two biological reserves: La Cusinga and Oro Verde (10 sample points at each refuge). La Cusinga (250 ha) and Oro Verde (150 ha) are also private reserves, but without governmental incentives or protection. The height of the forest canopy in both the biological reserves and in the wildlife refuges was up to 35 m.
c. *Forest edge:* twenty sample points were located in areas along the edge of primary and old-secondary forests.
d. *Forest fallows (young secondary forest):* These were areas of forest regenerating in recently abandoned pasture lands (5-10 years after abandonment). Canopy height at these locations was up to 10-15 meters.

e. *Homegardens:* These were traditional Costa Rican homegardens composed of a combination of annual/perennial crops, ornamental plants, and timber and fruit trees.
f. *Agrosilvopastoral systems:* Most of the pastures on the region were used to raise beef cattle, and had been planted with exotic pasture species. All of the selected sites had either living fences or scattered trees that provided perches or food sources to birds.
g. *Tree plantations:* As explained prevoiusly, most tree plantations in the PTBC consist of teak (*Tectona* grandis), or amarillon *(Terminalia amazonia*). Teak is a valuable exotic timber species that is exported to Europe and the United States. Amarillon is native to Costa Rica and it is one of the most promising species for timber production in the country. Since we did not get access to enough number of *T. amazonia* plantations to set up 20 sample points, we located 10 points in teak and 10 in amarillon plantations so as to have a total of 20 points for this habitat type. Plantation size varied from 2 ha to 20 ha, teak plantations were the largest in size.
h. *Oil palm plantations*: In the northern region of the corridor, there were hundreds of oil palm plantations (*Elaeis guineensis*), which is an exotic crop to Costa Rica. We located the sample points from recent established plantations (1.5 m tall) to mature plantations (up to 10 m height).
i. *Villages*: We selected two main areas for sampling. The first one was the largest village or town within the corridor (Puerto Cortez). In Puerto Cortez, we located 10 samplepoints. The other 10 sample points were located in the small villages along the main highway.
j. *Residential tourism projects:* Because the Path of the Tapir is along the Central and South Pacific Coast of the country, there has been an exponential growth in the number of foreigners building summer or retirement houses in the area in the last 10 years. We located the points in areas where there were more than one house built.

Bird Surveys

Each point was surveyed three times in the summer and three times during the rainy seasons. At each sample point (in a 50m radius plot and during a period of 10 minutes) We recorded and identified every bird species that was seen, habitat type, tree or structure where it was found; as well as type of bird activity (e.g., foraging, nesting, moving, and perching). All surveys were conducted from sunrise to 9:15 am during clear days, avoiding windy situations. Field identification was conducted using the Manual of Birds of Costa Rica [60] and the Birds of Costa Rica [61], as well as an identification of their calls and songs.

Data Analysis

For each habitat type we calculated the total abundance and species richness of birds by combining data from the two sampling periods (summer and winter), as well as comparing between both periods. We also calculated Shannon diversity index, rarefaction curves, Sorensen similarity index, and the number of shared species using EstimateS 5.0 [62]. Rarefaction curves were plotted against the number of individuals observed, using the mean of the four commonly employed abundance-based estimators (ACE, CHAO1, JACK1, and BOOSTRAP) [63]. Species were assigned dietary guilds and habitat guilds following Stiles and Skutch [60]. We compared abundance, species richness, diversity, and habitat and feeding guilds among the ten habitat types, using analysis of variance and/or Kruskal Wallis tests. We also used cluster analysis to group habitat types based on species composition. All statistical analyses were conducted in PAST v 1.90 [64].

Chapter 4

RESULTS

BIRD ABUNDANCE AND DIVERSITY IN THE PATH OF THE TAPIR BIOLOGICAL CORRIDOR

A total of 21,015 birds which included 293 species were identified during 200 hours in the ten land-use types (Appendix 1). From the observed birds, 11,045 individuals (271 species accumulated) corresponded to the summer surveys and 9,970 individuals (246 species) corresponded to winter surveys (Figure 2). Even though there was a difference of 25 species between both surveys, it is important to highlight that there were 47 species that were only sighted in the summer surveys and not in the winter season, and 22 species found only during the winter surveys. From the 47 species found in summer, 35 species (75%) corresponded to migratory species. On the contrary, almost 100% of the species registered in winter corresponded to resident species that are considered rare or uncommon species.

The most commonly observed species were (percentage from the total) *Ramphocelus costarricensis* (1064 individuals – 5.0%), *Thraupis episcopus* (754 individuals – 3.6%), *Quiscalus mexicanus* (743 – 3.5%), *Sporophila americana* (719 – 3.4%), *Troglodytes aedon* (702 – 3.3%), *Pitangus sulphuratus* (686 - 3.2%), and *Volantinia jacarina* (618 – 2.9%). Contrary to the abundance of these species, there were 121 species (41%) that had least than 10 individuals, and from this group, 23 species were sighted just once.

The majority of birds were insectivorous, accounting for 162 species and 42.4% of all observations. Frugivorous (56 species, 34.1%) and Granivorous (23 species, 13.9%) were the second and third most common guilds, respectively. Nectarivorous, Omnivorous, Carnivorous, and Piscivorous accounted by least than 10% of the observations all together.

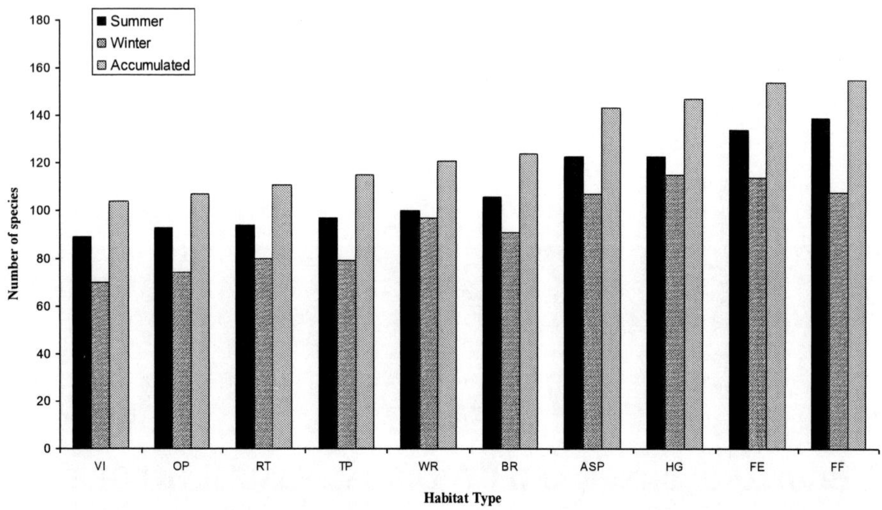

Figure 2. Accumulated number of bird species found in ten different habitat types in the Path of the Tapir Biological Corridor, Costa Rica.

In addition, most registered birds were classified as open area specialist (55 species, 40%), followed by general species (100 species, 32.5%), and forest specialist (123 species, 25.6%).

Comparison Across Sites in Bird Abundance and Diversity

There were differences in the mean abundance, species richness, and diversity of birds observed across the ten land-use types (Table 1). Agrosilvopastoral systems and homegardens were the most bird-abundant habitats, followed by the villages. On contrary, most forested habitats (tree plantations, biological reserves, wildlife refuges, and forest edges) were the least abundant ones. While forest fallows, forest edges, agrosilvopastoral systems, homegardens, and the reserves shared the highest species richness and diversity of birds, tree plantations and some open area habitats such as villages, oil palm plantations, and residential tourism projects had the lowest diversity values.

Rarefaction curves of the accumulated number of bird species confirmed some of the patterns described earlier (Figure 3). Forest fallows and forest edges had the highest expected number of species, followed by the wildlife refuges, biological reserves, and tree plantations.

Table 1. Abundance, species richness, and diversity of birds registered in ten habitat types in the Path of the Tapir Biological Corridor, Costa Rica

Habitat Type	Abundance (n/census)	Species Richness (species/census)	Shannon Index
Homegardens	562 (61.5)a	82 (3.3)ab	3.78 (0.03)ab
Agrosilvopastoral Systems	516 (55.0)ab	79 (3.0)ab	3.79 (0.04)ab
Villages	480 (66.5)b	56 (2.8)bc	3.38 (0.04)c
Oil Palm	362 (21.4)c	55 (1.2)c	3.42 (0.03)c
Forest Fallows	349 (33.6)c	83 (5.6)a	3.85 (0.05)a
Residential Tourism	315 (34.7)c	59 (2.7)bc	3.63 (0.05)b
Forest Edges	261 (29.9)cd	77 (1.8)ab	3.83 (0.05)a
Wildlife Refuges	233 (19.8)cd	67 (1.4)bc	3.80 (0.03)ab
Biological Reserves	219 (31.2)cd	65 (2.3)bc	3.81 (0.05)ab
Tree Plantations	205 (38.3)d	57 (2.5)bc	3.70 (0.06)b

Data represents means per census (N = 6) and standard errors. Different letters within same column indicate differences across land use types.

The villages, oil palm, and residential tourism projects had the lowest expected number of species. The agrosilvopastoral systems and the homegardens had intermediate results between these two extremes.

Biological reserves, wildlife refuges, and forest edges had the most similar assemblage of species (Table 2, Figure 4). This may be due to the fact that these three habitats are dominated by primary and old-secondary forest, having a majority of forest-specialist bird species. Agrosilvopastoral systems, homegardens, residential tourism, oil palm plantations were also very similar among themselves since they were dominated by general and open-area bird species. Tree plantations and forest fallows were also very similar because they shared a high number of both general and forest-specialist bird species. As expected, habitats that did not share many species also differed sharply in their proportion of forest cover. For instance, biological reserves and wildlife refuges were dominated by forest-specialist species while villages had a majority of open-area and general bird species.

From the 293 species observed, there were only 13 species that were observed in the ten habitat types and 16 species that were found in 9 land-use types, confirming the highly distinct bird assemblages in the study area.

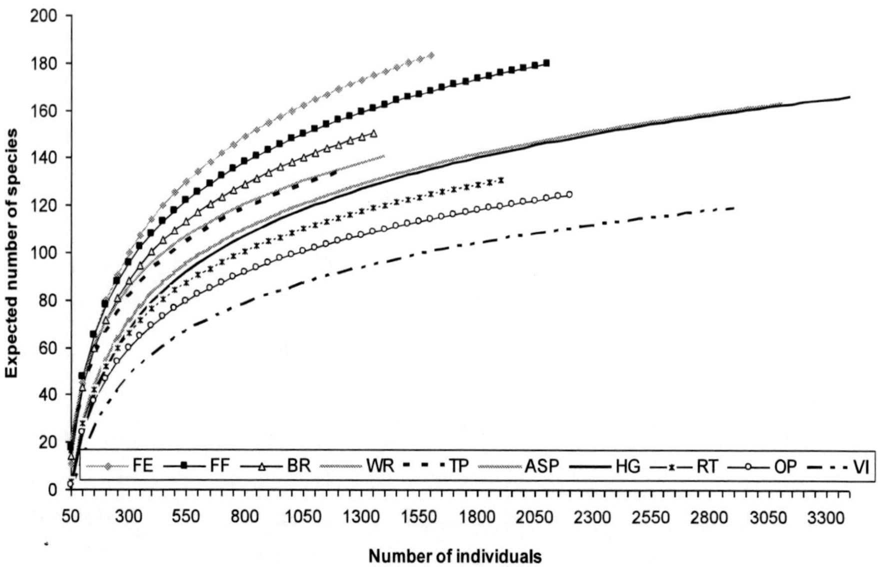

Figure 3. Rarefaction curves for the accumulated data of bird species in PTBC.

Table 2. Sorensen similarity index and shared species of ten habitat types in the Path of the Tapir Biological Corridor, Costa Rica

Habitat 1	Habitat 2	Species 1	Species 2	Shared species	Sorensen Index
BR	WR	124	121	102	0.833
HG	RT	148	111	102	0.788
HG	ASP	148	143	110	0.756
FF	TP	155	115	102	0.756
OP	RT	107	111	82	0.752
ASP	RT	143	111	93	0.732
VI	RT	104	111	78	0.726
VI	TP	111	115	81	0.717
HG	VI	148	104	90	0.714
HG	OP	148	155	108	0.713
FF	RT	155	111	94	0.707
FE	WR	154	121	96	0.698
OP	ASP	107	143	87	0.696
FE	FF	154	155	107	0.693

Habitat 1	Habitat 2	Species 1	Species 2	Shared species	Sorensen Index
HG	OP	148	107	88	0.69
VI	ASP	104	143	84	0.68
VI	OP	104	107	71	0.673
BR	FE	124	154	93	0.669
HG	TP	148	115	87	0.662
FE	TP	154	115	88	0.654
FF	ASP	155	143	95	0.638
OP	TP	107	115	68	0.613
ASP	TP	143	115	77	0.597
FF	OP	155	107	75	0.573
VI	FF	104	155	74	0.571
FE	RT	154	111	75	0.566
VI	TP	104	115	61	0.557
HG	FE	148	154	84	0.556
BR	FF	124	155	73	0.523
FF	WR	155	121	72	0.522
BR	TP	124	115	61	0.51
WR	TP	121	115	60	0.508
FE	ASP	154	143	71	0.478
FE	OP	154	107	58	0.444
VI	FE	104	154	54	0.419
WR	RT	121	111	48	0.414
BR	RT	124	111	46	0.391
HG	BR	148	124	51	0.375
HG	WR	148	121	49	0.364
ASP	WR	143	121	40	0.303
BR	ASP	124	143	40	0.3
OP	WR	107	121	33	0.289
BR	OP	124	107	31	0.268
VI	WR	104	121	26	0.231
BR	VI	124	104	26	0.228

BR = Biological Reserves, FE = Forest Edges, FF = Forest Fallows, HG = Homegardens, OP = Oil palm, ASP = Agrosilvopastoral systems, WR = Wildlife Refuges, RT = Residential Tourism Projects, TP = Tree Plantations, VI = Villages.

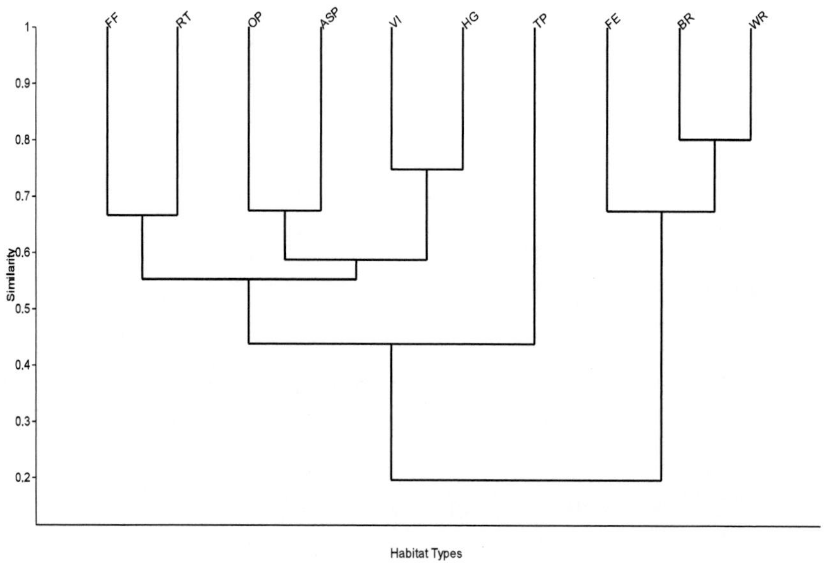

BR = Biological Reserves, FE = Forest Edges, FF = Forest Fallows, HG = Homegardens, OP = Oil palm, ASP = Agrosilvopastoral systems, WR = Wildlife Refuges, RT = Residential Tourism Projects, TP = Tree Plantations, VI = Villages.

Figure 4. Similarity of bird assemblages across ten habitat types in the Path of the Tapir Biological Corridor, Costa Rica

From the 13 species reported in the ten habitats, four corresponded to migratory species: *Legatus leucophaius, Vermivora peregrine, Dendroica pensylvanica,* and *Catharus ustulatus*; and nine were native species: *Ramphastos swainsonii, Pionus senilis, Leptotila verreauxi, Tangara larvata, Tityra semifasciata, Leptotila cassini, Cyanocompsa cyanoides*, and *Attila spadiceus*. Contrary to the capability of adaptation of these species to different habitat types, there were 47 species that were registered in only a single habitat type: 11 in agrosilvopastoral systems, 7 in forest edges, 6 in biological reserves, 5 in oil palm plantations, 5 in villages, 4 in homegardens, 4 in forest fallow, 2 in wildlife refuges, 1 in tree plantations, and none in the residential tourism projects. Furthermore, 25 species were only observed in either or both the biological reserves and wildlife refuges, and 21 species were only registered in the agroforestry systems.

SPECIES OF CONSERVATION CONCERN

Bird species of conservation concern are those more sensitive to be extirpated from a site following habitat loss, species with larger or heavier bodies and those foraging on insects, fruits, or both [65]. According to the Law 7317 for the Conservation of Wildlife in Costa Rica, at present, there are 80 bird species of conservation concern and 17 that are considered endangered species in the country.

We registered a total of 26 species of conservation concern and three endangered species (Table 3). Most species (20 out of 29) were forest specialist, followed by general (5 species) and open area species (4 species). Moreover, 55.2% (16 species) were frugivorous, 24.1% (6 species) insectivorous, 13.8% carnivorous (4 species), and 3.4% (1 species) nectarivorous. Overall, all land-use types provided habitat to at least one of the species of conservation concern, and a few of the bird species were observed in more than one habitat. For instance, forest edges, biological reserves or wildlife refuges harbored 18 of the species of concern; and the two agroforestry systems surveyed in this study provided habitat for 13 of these species, including the *Ara macaw*, one of the three endangered species found in this study.

Table 3. Species of conservation concern found in ten habitat types in the Tapir of the Tapir Biological Corridor, Costa Rica

Common name	Scientific name	Family	Guilds
Reduced populations			
Great Black-Hawk	*Buteogallus urubitinga*	Accipitridae	F, C
King Vulture	*Sarcoramphus papa*	Cathartidae	F, C
Scaled Pigeon	*Patagioenas speciosa*	Columbidae	G, Fr
Turquoise Cotinga	*Cotinga ridgwayi*	Cotingidae	F, Fr
Yellow-billed Cotinga	*Carpodectes antoniae*	Cotingidae	F, Fr
Crested Guan	*Penelope purpurascens*	Cracidae	F, Fr
Great Curassow	*Crax rubra*	Cracidae	F, Fr
Wedge-tailed Grass-Finch	*Emberizoides herbicola*	Emberizidae	OA, Gr
Bat Falcon	*Falco rufigularis*	Falconidae	OA, C
Collared Forest-Falcon	*Micrastur semitorquatus*	Falconidae	F, C

Table 3. (Continued)

Common name	Scientific name	Family	Guilds
Long-tailed Woodcreeper	*Deconychura longicauda*	Furnariidae	F, I
Yellow-tailed Oriole	*Icterus mesomelas*	Icteridae	G, I
Speckled Mourner	*Laniocera rufescens*	Incertae Sedis	F, I
Marbled Wood-Quail	*Odontophorus gujanensis*	Odontophoridae	F, I
Golden-naped Woodpecker	*Melanerpes chrysauchen*	Picidae	F, I
Blue-headed Parrot	*Pionus menstruus*	Psittacidae	G, Fr
Brown-hooded Parrot	*Pionopsitta haematotis*	Psittacidae	F, Fr
Crimson-fronted Parakeet	*Aratinga finschi*	Psittacidae	OA, Fr
Mealy Parrot	*Amazona farinosa*	Psittacidae	F, Fr
Orange-chinned Parakeet	*Brotogeris jugalaris*	Psittacidae	OA, Fr
Orange-fronted Parakeet	*Aratinga canicularis*	Psittacidae	G, Fr
Red-lored Parrot	*Amazona autumnalis*	Psittacidae	F, Fr
White-crowned Parrot	*Pionus senilis*	Psittacidae	G, Fr
Crested Owl	*Lophostrix cristata*	Strigidae	F, I
Great Tinamou	*Tinamus major*	Tinamidae	F, Fr
Baird's Trogon	*Trogon bairdii*	Trogonidae	F, Fr
Endangered species			
Mangrove Hummingbird	*Amazilia boucardi*	Trochilidae	F, N
Scarlet Macaw	*Ara macao*	Psittacidae	F, Fr
Red-crowned Ant-Tanager	*Habia rubica*	Thraupidae	F, I

Guilds: F = forest specialist, G = general, OA = open area species, Fr = Frugivorous, I = Insectivorous, N = nectarivorous, Gr = granivorous, C = carnivorous.

Chapter 5

DISCUSSION

CONTRIBUTION OF THE DIFFERENT LAND-USE TYPES TO BIRD DIVERSITY IN THE PTBC

The accumulated number of bird species observed in this study represented 34% of the avifauna registered in the whole country (293 out of 854 bird species in Costa Rica, [61]), and almost 62% of the species recorded in the Annual Christmas Count conducted in a part of the biological corridor in 2008 [Urena, N., personal communication, 2009]. However, we can assume that this number would be larger if the present study had included flyover species, wetland habitats, and the part of the corridor above 500 meters above sea level.

The persistence of tropical forest species in human-dominated landscapes is a fundamental challenge of tropical ecology and conservation [66]. Conservation in the PTBC is undertaken through 30 private reserves and the goal is to provide connectivity among these reserves [58]. Therefore, the contribution of agroforestry systems and other human-modified land-use types (matrix) may determine the success of the corridor in its role as connector among habitats. In the present study, the evaluated landscape matrix, including agrosilvopastoral systems, homegardens, tree plantations, oil palm plantations, villages, residential tourism projects, and the young secondary forest, harbored a total of 239 (81.5% of the observed species), highlighting its relevance for bird conservation in the corridor. However, the main forested habitats (wildlife refuges, biological reserves, and forest edges) harbored 66% of the species in the area, and were also the main habitat for species of conservation concern [24, 25, 29, 67, 68].

Contribution of Agroforestry Systems to Bird Conservation in the Corridor

Agroforestry systems are becoming increasingly important land uses in the tropics due to their functions in improving food security, contributing to rural development and poverty alleviation, and providing several environmental services such as carbon sequestration and biodiversity [69 - 71]. The contribution of agroforestry systems to biodiversity conservation has been studied by several authors [71 - 75]. Overall, some studies have shown that certain types of agroforestry systems– such as windbreaks, agrosilvopastoral systems, homegardens, and perennial crop-tree combinations–can contain bird assemblages that are as abundant, species-rich and diverse as those in natural forests of the same region [75]. However management type and intensity of management practices, including fertilization, herb control, mulching, and shade, can have potentially different effects on biodiversity [76].

In addition, wildlife diversity can vary depending on wildlife species and guilds. For example, several authors have found that forest-specialist bird species tend to be less abundant, while open-area species, granivores, and general species populations tend to be larger in agroforestry systems than in nearby forests [71, 75]. In this study, forested habitats (wildlife refuges, biological reserves) and agroforestry systems (homegardens, agrosilvopastoral systems) shared less than 51 bird species, thus showing this difference in species composition. Biodiversity conservation in agroforestry systems can increase if there is diversification of native species, and other plants that ensure structural complexity that provide fruits and other resources to wildlife [75].

Role of Agrosilvopastoral Systems in Bird Species Conservation and Potential for Improvement

According to recent statistics, in the Neotropics pastures and agrosilvopastoral land represent 77% of the total of 500 million ha of agricultural land [77]. About 30% of these pastures are on poor acid soils [78], as is the case in the present study. Agrosilvopastoral systems that involve the combination of trees with pastures and livestock offer an alternative set of cattle production systems. These systems are more complex than grass monocultures and are classified based on the functions and configuration or structure of trees within the system; examples are dispersed trees in pastures; live fences in pastures; fodder banks; tree alley pasture systems; and pastures with windbreaks [79]. Agrosilvopastoral systems can provide benefits to cattle farmers by enhancing nutrient cycling, increasing the sustainability of pastures, fodder production for animals, and diversification of income (income from timber, firewood, etc.) that contributes to the economic sustainability of cattle farms [80]. In addition, due to their increased complexity relative to grass monoculture systems, agrosilvopastoral practices also have important benefits for biodiversity [81 - 84].

For example, in many tropical pastures there are isolated trees that can be remnants of the original forest. They are conserved by farmers for their functions as shade for animals and humans, and as sources of firewood, timber and fruit. These trees and shrubs may be important in attracting birds, in comparison with tree-less grass monocultures. Crown size and shape, tree height, fruit production and other characteristics may influence the ability of trees to attract birds [85, 86]. Apparently birds prefer taller trees as perches to have a better view of the surroundings and to avert potential predators. On the other hand, results of a recent study in the Pacific region of Costa Rica suggested that crown diameter was more important than height of trees in attracting birds [84].

As shown earlier, a combination of several characteristics may determine the role of each tree species to attract birds to trees in agrosilvopastoral systems. Distance to forest patches and position in the matrix can also have an important influence. For example, to evaluate the contribution of living fences to improving the connectivity of the agricultural matrix, Francesconi *et al.* [84] examined bird species composition (forest specialists, open-area specialists, and general species) that use living fences as habitat in Esparza, located in the Pacific lowlands of Costa Rica. Bird species composition changed as a function of distance to the forest patch; yet species richness, number of individual birds and Shannon Diversity indices were similar between forest interiors (control) and living fences. Their results suggest that living fences near forest patches provide habitat to many bird species including forest specialists. However, fence structure and composition significantly influenced usage by birds. The authors concluded that the presence of diverse native tree species in fences and increased vegetative cover may counteract the effect of distance to the forest patch, promoting greater bird species diversity in living fences and in the landscape [84].

As seen, the specific types of agrosilvopastoral practices adopted by farmers are important for the conservation of biodiversity [87, 88]. In a long term project in landscapes dominated by cattle in Esparza (Costa Rica), Matiguas (Nicaragua), and Quindio (Colombia), intensive monitoring of birds in different land use systems over four years showed that the agrosilvopastoral practices with high tree densities in pastures and multi-strata or permanent live fences had significantly higher abundance and species richness of birds than degraded and grass monoculture pastures, with comparable values to those of forested vegetation [89]. In their study, Ibrahim *et al.* [89] reported that a total of 111 bird species were observed in the landscape of Esparza, Costa Rica, 170 species in Quindio, Colombia and 154 species in Matiguas, Nicaragua. Of these, a total of 60.5%, 54% and 64% of the species inventoried were dependent on the forest in Costa Rica, Colombia, and Nicaragua, respectively. Likewise, in a fragmented agricultural landscape in Cañas, Costa Rica, Cardenas *et al.* [90] showed that the agrosilvopastoral systems with high tree densities and permanent live fences (multi-strata fences) had more than 40 bird species, whereas grass monoculture pastures had 28 bird species.

These data indicate that it is possible to implement agrosilvopastoral systems that are compatible with production objectives and biodiversity conservation. Good farm planning is needed including a mosaic of land uses that are compatible with both the conservation of biodiversity and production objectives [83, 88]. For example, cattle farms should incorporate land uses which include establishment of fodder banks, multi-strata live fences, and pastures with high tree cover or density, as well as riparian and secondary forest [89].

The agrosilvopastoral systems evaluated as part of this study were characterized by the presence of native and exotic tree/shrub species that were either sparse in the pasture fields as remnant trees, or were part of living fences. This vegetation may play an important role for biodiversity conservation within the corridor because it provides habitat, resources, and facilitates connectivity that is otherwise absent in the agricultural landscape of the region [81, 90, 91]. Therefore, diversifying agrosilvopastoral systems through the use of native species may be a priority to enhance connectivity along the corridor. This may be especially important because more than 20% of the PTBC is composed of open pasture lands [73].

ROLE OF HOMEGARDENS IN CONSERVATION OF AVIAN DIVERSITY IN THE PTBC

The homegardens provide the household with a basic food source as well as high value products to generate cash are important in Latin America, and are often used as tools in development projects that promote food security, especially in the poorest areas [92, 93]. In several regions of Costa Rica homegardens are important for supplying food [94]; they also serve as a buffer in times of harvest failure or economic depression [95].

As in other regions worldwide, the structure and composition of homegardens in Latin America are quite complex: they are quite diverse in vertical and horizontal structure and well as in plant species composition. Most homegardens of Latin America consist of several vertical and horizontal strata in which plants are arranged according to their adaptability to the existing light conditions and nutrient resources. The number of individual plants per stratum varies among homegardens; older, more mature homegardens display more developed tree strata. Some homegardens resemble agricultural fields with an emphasis on the herbaceous and low shrub strata, with a greater focus on agricultural crop production. Others have more trees, with architecture similar to that of the native forest of the region [95 - 97].

Plant species composition in homegardens is influenced by access to water, owners' economic activities, labor availability, traditional social organization, modernization processes, and economic development [96, 98]. Both exotic and native plants are used, with emphasis on fruit trees. Locally, plant species diversity of homegardens can be influenced by size of the homegardens. For example, in Nicoya, Costa Rica, Lok *et al.* [97] found that the size of homegardens ranged from 0.1 to 1.4 ha with an average of 0.5 ha. The smallest homegardens considered in the study had the highest plant species diversity, with 205 to 745 species and an average of 348 species per ha. In contrast, the larger homegardens had only an average of 96 species per ha, with less variability among homegardens, in comparison with the smaller homegardens that exhibited higher variability in their species diversity.

Homegardens may have positive effects on biodiversity of wildlife, as they can serve as local refuges for plants and animals that otherwise may be threatened by human or natural disturbances [93]. For example, Griffith [99] reported that during the 1998 fires in Petén, Guatemala, homegardens and other agroforestry systems may have served as critical refuge during a habitat bottleneck for many forest species. Apparently, agroforestry farms attracted birds by virtue of their complex structure – similar to that of intact forest patches – as they harbor insects, provide nesting sites, and offer protection from predators [99]. They were also attracted by the cultivated fruit trees, which may have provided some of the only food sources in the region after fire destroyed most of the surrounding vegetation.

Homegardens are important for bird conservation because their large plant diversity tends to attract many bird species of different guilds [71]. In spite of their many advantages in ensuring food supply and other benefits to local people, currently in the PTBC homegardens are gradually disappearing as they are displaced by increasing urbanization and tourism development [58]. Increased urbanization typically leads to reduction in species richness [85]. For instance, in this study there was decrease in bird species richness from 148 species in homegardens, to only 111 and 104 species in residential tourism projects and in villages, respectively.

In this study, the evaluated homegardens were located in areas with unpaved roads and houses of small size in relation to the garden. In contrast, point counts in villages were located along paved roads, and most of the landscape was composed of cement structures, surrounded by some native and exotic vegetation. Residential tourism projects were located either in fragmented forest landscapes, or they were surrounded by pasture lands. In addition, in these sites houses were of large size in comparison to those in the homegarden and village sites. Therefore, because birds respond to vegetation composition and structure [75, 100, 101], it may be recommended to retain more native vegetation in villages and residential projects, as well as to reduce the size of the urbanized areas so as to retain more bird diversity [100].

Contribution of Secondary Forests and Tree Plantations to Bird Conservation

Secondary forests and tree plantations are rapidly expanding across tropical landscapes [63]. Secondary forests and tree plantations may play an important role to biodiversity conservation by increasing regional-scale species richness at the landscape level [24, 63]; improving connectivity across the landscape, and buffering existing forest fragments [102]. In the current study, forest fallows and tree plantations shared less than 73 species with either the biological reserves or wildlife refuges. Forest fallows provided habitat to 53 forest species and the tree plantations to 42 species. Even though forest fallows can provide habitat for many bird species, regaining the complex microhabitats and structures required by primary forest specialist is likely to take centuries [78, 103]. It is likely that the older the secondary forest the more forest species may inhabit the area. Therefore, allowing these forest fallows to re-grow by themselves may increase the connectivity between forested reserves in the corridor.

It has been observed that tree plantations promote the regeneration of understory species by shading out grasses, increasing nutrient richness of topsoil, allowing the growth of more sensitive tree species, and creating a microclimate that attracts seed dispersers [104, 105]. In addition, tree plantations may serve as perches, food sources and habitat for a number of wildlife species, thus contributing to increase overall biodiversity [85, 86, 106].

Use of native species, especially in mixtures, may increase wildlife populations in former degraded landscapes such as those currently present in the PTBC [107]. In this study, we found 96 bird species in the native *Terminalia amazonia* plantations and only 71 in the exotic *Tectona grandis* plantations. Although a formal statistical test was not run to compare plantations of the two species due to lack of enough sites to allow for 20 point counts in each (as explained in the Methods section), this appears to be an interesting trend. However, management intensity could also have contributed to these differences in bird diversity among the two species [76]. For example, all understory vegetation growing in the *T. grandis* was removed, as it is commonly done in teak plantations in Costa Rica and elsewhere in Latin America. In contrast, *Terminalia amazonia* plantations were less intensively managed than *T. grandis* plantations, possibly due to the lack of well defined silvicultural guidelines for *Terminalia.* In the *T. amazonia* plantations visited in this study, most of the natural regenerated vegetation was not removed and thus remained as a part of the natural landscape. Recent studies conducted in experimental plantations of several native species in the Caribbean region of Costa Rica have also found abundant and species-diverse understory growth in unmanaged *T. amazonia* plantations [108, 109]. This understory may provide habitat and other resources to bird and other wildlife species [106, 108].

To increase bird and other wildlife abundance and diversity in the PTBC, it may be wise to encourage and promote the establishment of plantations of native species such as *T. amazonia,* instead of those of exotics such as teak or oil palm. *Terminalia amazonia* is considered a very promising native timber species for plantation in Central America due to high value and good growth, and it is already one of the most frequently planted native species by local farmers in the Caribbean region of Costa Rica [110, 111]. In Panama, the commercial company Futuro Forestal reported sales of first thinnings of young *T. amazonia* plantations from their farms in Chiriqui that achieved the same price at *T. grandis*, due in part to Forest Stewardship Council certification [112]. *T. amazonia* is apparently capable of adapting to different site conditions, with good growth even on degraded sites covered by invasive vegetation [113]. Other native tree species could be planted with similar advantages. Improved knowledge and application of management techniques (thinning, pruning) and genetic selection can also contribute to improved performance [114]. In Costa Rica, incentives such as Payments for Environmental Services (PES) are available to establish tree plantations including native species.

Contribution of Habitat Types of the PTBC to Species of Conservation Concern

In order to evaluate the role of different habitat types to biodiversity conservation, it is important to know not only how many species are present within these habitats, but also which species are present and whether any of these are of conservation concern [75].

The nine species of the Psittacidae family registered in this study are considered of conservation concern or endangered (e.g. *Ara* macaw) as a result of habitat loss, direct hunting, and poaching for the pet trade market [60, 61]. However, most of these species were observed in more than one habitat type and the majority of them had important abundance. Of more concern are *Pionopsitta haematotis,* which individuals were only found in private reserves, and *Ara macaw*, which once resided on both the Pacific and Caribbean slopes, and now is only located in a few spots along the Pacific Coast [60, 61, 114]. In the corridor, *Ara macaw* populations were located foraging and perching mostly in homegardens, villages, and in forest edges in the southern part of the region. A few individuals were also registered in homegardens of the PTBC's northern limit; however, according to a few neighbors these birds were part of a re-introduction program back in the 1990s.

The other two endangered species *Habia rubica* and *Amazilia boucardi* were registered only once over the time of this project. *Habia rubica* is a forest-dependent species that forages in the understory of mature forests and *Amazilia boucardi* is an endemic species of the Pacific coast of Costa Rica, specifically mangrove and adjacent forests [61]. Another group of species that also depend on the conservation of forest for their survivorship in the corridor are *Buteogallus urubitinga*, *Sarcoramphus papa, Cotinga ridgwayi, Carpodectes antoniae, Penelope purpurascens*, *Crax rubra*, *Micastur semitorquatus*, *Deconychura longicauda*, *Laniocera rufescens*, *Odontophorus gujanensis*, *Pionopsitta haematotis*, *Lophostrix cristata*, *Tinamus major*, and *Trogon bairdii.* Unfortunately, urban and tourism development are jeopardizing the conservation of many of these species in the region, especially because forest cover is either disappearing or being fragmented around most private reserves.

Chapter 6

Conclusions

As highlighted recently by several researchers, a new conservation paradigm must incorporate human-modified landscapes in assessments of biodiversity and ecosystem services, planning of corridors, establishment of buffer zones, and restoration of degraded lands [115]. The current study aimed to incorporate this approach, thus we assessed biodiversity not only in private forest reserves, but also in several other habitat types that were part of this human-modified biological corridor known as Path of the Tapir.

Agroforestry systems such as agrosilvopastoral systems and homegardens are diverse habitats that if managed in the proper way may increase the current contribution to the PTBC and other fragmented habitats to wildlife conservation. Adding native tree species, especially fruit trees, in agricultural systems adjacent to forested areas may add conservation value to these habitat types, attracting bird species such as *Ara macaw* and other forest-dependent bird species.

Forest fragmentation, unplanned urbanization, monocultures, and the introduction of exotic plant species should be avoided in biological corridors in order to not jeopardize biodiversity. One example of these monocultures that should be avoided is oil palm plantations, which is exotic to Costa Rica. Even though there were found more than 100 species in the oil plant plantations, their species richness value was one of the lowest of all evaluated habitats. Furthermore, the worldwide expansion of this type of plantation is cited as a major threat to tropical biodiversity [116, 117].

In contrast, planting native timber tree species, such as *Terminalia amazonia* and other native tree species of interest in Costa Rica, and elsewhere, should be given priority to serve biodiversity and economic objectives of humans inhabiting the corridor.

At the species level, aside from the 29 species of conservation concern, special priority should be given to favoring habitat types preferred by the forest-specialist species. This is especially important with regard to frugivorous and insectivorous birds that are more sensitive to human disturbances and therefore more prone to extinction.

Finally, we need to stress that biodiversity conservation should always include primary and old-secondary forests (e.g., those forests located in wildlife refuges, biological reserves, or isolated forest fragments) as these are the most important habitat type for wildlife at the landscape level. However, the results of this research also show that the matrix, especially agroforestry systems and plantations, are important to ensure wildlife conservation in the biological corridor in the short and long term.

Appendix 1. List of Species Registered in the Different Habitat Types in the Path of the Tapir Biological Corridor, Costa Rica

Species	HG	BR	VI	FE	FF	OP	ASP	WR	RT	TP	Total
Acadian Flycatcher				1						1	2
Alder Flycatcher					2		3				5
Amazon Kingfisher			1	1							2
American Redstart		1						1			2
Baird's Trogon		5						4			9
Baltimore Oriole	14		17		15	8	21		8	7	90
Bananaquit	42		13	1	23		26		23		128
Band-tailed Barbthroat		3		9	1			4		3	20
Bank Swallow			7			3			1		11
Bare-crowned Antbird		4		3				4			11
Bare-throated Tiger-Heron							3				3
Barn Swallow	1										1
Barred Antshrike	8	2	1	9	26	2	9		12	12	81
Bat Falcon							1				1
Bay-breasted Warbler					1						1
Bay-headed Tanager	5	2		1	8		5				21
Bicolored Antbird		1						2			3
Black Vulture	13		44			5	10		2	5	79
Black-and-white Warbler	2			2	2					1	7
Black-bellied Whistling-Duck						33	2				35
Black-bellied Wren				11	5				2		18
Black-cowled Oriole						2	1				3

Appendix 1. (Continued)

Species	HG	BR	VI	FE	FF	OP	ASP	WR	RT	TP	Total
Black-crowned Tityra	19		4	6	4	11	16		6	3	69
Black-faced Antthrush		1		2				3			6
Black-hooded Antshrike	3	62		76	60	2		81	6	26	316
Black-striped Sparrow	22	1	4		65	63	70		19	6	250
Black-striped Woodcreeper		14		8				16		4	42
Black-tailed Flycatcher		4		6				3			13
Black-throated Trogon		10		8				18			36
Blue Dacnis	10		6	7	2		8	2	5		40
Blue Ground-Dove		2		3	3			2			10
Blue-and-white Swallow	2						7				9
Blue-black Grassquit	54		45		51	180	223		65		618
Blue-black Grosbeak	2	8	2	8	12	4	2	11	1	3	53
Blue-crowned Manakin		68		58	6			60		9	201
Blue-crowned Motmot	12	21		22	18		10	16	7	7	113
Blue-gray Tanager	181		192		83	37	113		98	50	754
Blue-headed Parrot	2		6								8
Blue-throated Goldentail	7	8	3	16	25	10	4	10	4	29	116
Boat-billed Flycatcher	13		11	3	9	9	16	2	11	15	89
Bright-rumped Attila	7	8	2	5	5	2	6	1	3	12	51
Broad-winged Hawk		1		2				2	1		6
Bronzed Cowbird	5		3			24	33				65
Bronzy Hermit				9				1			10
Brown Jay	7		4	8	11	25	70		7	31	163

Appendix 1. List of Species Registered in the Different Habitat Types in the Path of the Tapir Biological Corridor, Costa Rica

Species	HG	BR	VI	FE	FF	OP	ASP	WR	RT	TP	Total
Acadian Flycatcher				1						1	2
Alder Flycatcher					2		3				5
Amazon Kingfisher			1	1							2
American Redstart		1						1			2
Baird's Trogon		5						4			9
Baltimore Oriole	14		17		15	8	21		8	7	90
Bananaquit	42		13	1	23		26		23		128
Band-tailed Barbthroat		3		9	1			4		3	20
Bank Swallow			7			3			1		11
Bare-crowned Antbird		4		3				4			11
Bare-throated Tiger-Heron							3				3
Barn Swallow	1										1
Barred Antshrike	8	2	1	9	26	2	9		12	12	81
Bat Falcon							1				1
Bay-breasted Warbler					1						1
Bay-headed Tanager	5	2		1	8		5				21
Bicolored Antbird		1						2			3
Black Vulture	13		44			5	10		2	5	79
Black-and-white Warbler	2			2	2					1	7
Black-bellied Whistling-Duck						33	2				35
Black-bellied Wren				11	5				2		18
Black-cowled Oriole						2	1				3

Appendix 1. (Continued)

Species	HG	BR	VI	FE	FF	OP	ASP	WR	RT	TP	Total
Black-crowned Tityra	19		4	6	4	11	16		6	3	69
Black-faced Antthrush		1		2				3			6
Black-hooded Antshrike	3	62		76	60	2		81	6	26	316
Black-striped Sparrow	22	1	4		65	63	70		19	6	250
Black-striped Woodcreeper		14		8				16		4	42
Black-tailed Flycatcher		4		6				3			13
Black-throated Trogon		10		8				18			36
Blue Dacnis	10		6	7	2		8	2	5		40
Blue Ground-Dove		2		3	3			2			10
Blue-and-white Swallow	2						7				9
Blue-black Grassquit	54		45		51	180	223		65		618
Blue-black Grosbeak	2	8	2	8	12	4	2	11	1	3	53
Blue-crowned Manakin		68		58	6			60		9	201
Blue-crowned Motmot	12	21		22	18		10	16	7	7	113
Blue-gray Tanager	181		192		83	37	113		98	50	754
Blue-headed Parrot	2		6								8
Blue-throated Goldentail	7	8	3	16	25	10	4	10	4	29	116
Boat-billed Flycatcher	13		11	3	9	9	16	2	11	15	89
Bright-rumped Attila	7	8	2	5	5	2	6	1	3	12	51
Broad-winged Hawk		1		2				2	1		6
Bronzed Cowbird	5		3			24	33				65
Bronzy Hermit				9				1			10
Brown Jay	7		4	8	11	25	70		7	31	163

Species	HG	BR	VI	FE	FF	OP	ASP	WR	RT	TP	Total
Brown-chested Martin			2								2
Brown-hooded Parrot		5		4				13			22
Buff-rumped Warbler		8	1	2				1			12
Buff-throated Foliage-gleaner		1						2			3
Buff-throated Saltator	12		1	8	19		12		9	4	65
Cattle Egret	7		5			34	174		1	5	226
Charming Hummingbird	4			4	1						9
Cherries Tanager	218		169	24	222	75	140		169	47	1064
Chesnut-backed Antbird	3	97		100	45	4		134	2	30	415
Chesnut-mandibled Toucan	47	71	5	97	38	4	27	80	55	30	454
Chesnut-sided Warbler	6	8	5	23	28	2	22	2	6	31	133
Clay-colored Robin	123		103	2	23	5	23		43		322
Cocoa Woodcreeper	18	26		42	21	10	6	31	7	23	184
Collared Forest-Falcon		2						1			3
Common Pauraque		1			2		1	1		1	6
Common Tody-Flycatcher	15		17		6	2	18		6	1	65
Common Yellowthroat		1									1
Crested Caracara	1		1			12	8		1	3	26
Crested Guan		6						8			14
Crested Owl				5							5
Crimson-fronted Parakeet	24		17	5		4	27		2	2	81
Dickcissel							1				1
Dot-winged Antwren		16		7	2			11			36
Double-toothed Kite		3		3				6			12
Dusky Antbird		7		8	2			3			20

Appendix 1. (Continued)

Species	HG	BR	VI	FE	FF	OP	ASP	WR	RT	TP	Total
Dusky-capped Flycatcher				4	3	3	3	1	1	6	21
Eastern Meadowlark							2				2
Eastern Wood-Pewee	1						9				10
Eye-ringed Flatbill		2						2			4
Fasciated Tiger-Heron							1	1			2
Ferruginous Pygmy-Owl	2		2								4
Fiery-billed Aracari	34	28	6	51	20		18	29	37	22	245
Garden Emerald	6		3		5	1	3		3		21
Giant Cowbird						2	1				3
Golden-crowned Spadebill		8		2				21			31
Golden-hooded Tanager	47	6	31	27	53	2	48	7	38	23	282
Golden-naped Woodpecker	1	28		31	8		4	20	3	2	97
Golden-winged Warbler					1			1			2
Gray Hawk				1		.	1		1		3
Gray-breasted Martin	2		17				5		1		25
Gray-capped Flycatcher	18	1	15		4		10		7		55
Gray-chested Dove	7	7	15	8	6	5	4	5	2	3	62
Gray-crowned Yellowthroat							2				2
Gray-headed Chachalaca	32		5		41				21		99
Gray-headed Dove		1		2				1			4
Gray-headed Kite				1				1			2
Gray-headed Tanager	6	36		27	22			29		5	125
Gray-necked Wood-Rail	8	1	4	2	8	12	6	1	1		43
Great Antshrike		2		9	6					2	19

Species	HG	BR	VI	FE	FF	OP	ASP	WR	RT	TP	Total
Great Black-Hawk		1						1			2
Great Crested Flycatcher	10		3	4	14	9	1	2	12	6	61
Great Curassow		11		2				16			29
Great Egret	1					3	12				16
Great Kiskadee	131		157	1	54	114	128		74	27	686
Great Tinamou		17		6	6	1		24			54
Great-tailed Grackle	162		408		1	72	79		17	4	743
Green Heron			2				5				7
Green Honeycreeper	7	5		3	6		1	2	2		26
Green Kingfisher				3		1	1				5
Green Shrike-Vireo		4						2			6
Green-breasted Mango	7		1	1							9
Greenish Elaenia				2	1						3
Groove-billed Ani	3		2			63	12		8		88
Hoffmann's Woodpecker					1				2		3
House Sparrow			33								33
House Wren	164		138		18	184	73		90	35	702
Inca Dove	3		22			8	23		5		61
Indigo bunting	1										1
Kentucky Warbler					2					1	3
King Vulture				2						1	3
Laughing Falcon	2		1	3	4	2		6	1	4	23
Least Flycatcher					1					1	2
Lesser Elaenia	1						3				4

Appendix 1. (Continued)

Species	HG	BR	VI	FE	FF	OP	ASP	WR	RT	TP	Total
Lesser Greenlet		20		7	7	1	2	17	1	5	60
Lesser Swallow-tailed Swift			6								6
Lineated Woodpecker	3	1		2	3	6	7		3	12	37
Little Blue Heron						1	3				4
Little Tinamou		8		7	6			10		1	32
Long-billed Gnatwren		14		8	4			23		1	50
Long-billed Hermit	7	29		24	3	1		22	1	4	91
Long-billed Starthroat	4			5	3		1			3	16
Long-tailed Woodcreeper		1		1				2			4
MacGillivray's Warbler							1				1
Mangrove Cuckoo	1						1				2
Marbled Wood-Quail		8						2			10
Masked Tityra	18	2	8	13	8	3	10	2	1	1	66
Mealy Parrot	19	23	4	25		2	2	43	4	3	125
Melodious Blackbird	46		20		1	2	22		1	1	93
Mottled Owl		1									1
Mourning Warbler					3		2			1	6
Northern Barred-Woodcreeper		7		8	4			11		1	31
Northern Bentbill				1	2					1	4
Northern Jacana	2		7				13				22
Northern Rough-winged Swallow	5					4	11		2		22
Northern Waterthrush				1		1					2
Ochre-bellied Flycatcher		15		10	2			26		4	57
Olivaceous Piculet	3			3	8		4	5	2	3	28

Species	HG	BR	VI	FE	FF	OP	ASP	WR	RT	TP	Total
Olivaceous Woodcreeper										1	1
Olive-sided Flycatcher	1	1					3			1	6
Orange-billed Sparrow		13		3	1			6		3	26
Orange-chinned Parakeet	107		121	14	36	69	108		59	93	607
Orange-collared Manakin		7		3	10			3		3	26
Orange-fronted Parakeet	2										2
Orchard Oriole	4			1		1					6
Osprey			2								2
Pale-billed Woodpecker		12		16				27	1		56
Pale-breasted Ground-Dove	2										2
Pale-breasted Spinetail							20				20
Pale-vented Pigeon	50		46	1	2	3	15		13		130
Palm Tanager	168		138		76	45	64		103	10	604
Paltry Tyrannulet	1	1	1	1	4		6			2	16
Pearl Kite							2				2
Pheasant Cuckoo				1							1
Philadelphia Vireo	3		2	1	4	1	3		4	2	20
Piratic Flycatcher	57	20	62	26	26	14	56	15	40	23	339
Plain Antvireo		14		2	2			13			31
Plain Wren	30		13	3	32	51	26		37	10	202
Plain Xenops	1	6		4	7		3	14	3	4	42
Plain-breasted Ground-Dove	3		3								6
Purple Gallinule			8				5				13
Purple-crowned Fairy	12	5	1	9	11		1	4	13	5	61
Red-billed Pigeon	2						1				3

Appendix 1. (Continued)

Species	HG	BR	VI	FE	FF	OP	ASP	WR	RT	TP	Total
Red-capped Manakin		25		13	2			16			56
Red-crowned Ant-Tanager		2									2
Red-crowned Woodpecker	104	3	75	10	48	75	108		47	61	531
Red-eyed Vireo							2				2
Red-legged Honeycreeper	15	2		4	11		11	4	4		51
Red-lored Parrot	79	19	52	13	4		4	18	28	13	230
Red-rumped Woodpecker		1									1
Ringed Kingfisher			1		1						2
Riverside Wren	4	31		82	23	1		33	5	7	186
Roadside Hawk	16		13	1	24	32	39		19	38	182
Rock Pigeon			17								17
Rose-breasted Grosbeak				1	2						3
Rose-throated Becard		1		4	2			2		5	14
Royal Flycatcher				4							4
Ruby-throated Hummingbird	1				5						6
Ruddy Quail-Dove		5						7			12
Ruddy Woodcreeper		2						1			3
Ruddy-breasted Seedeater			2			3	5				10
Ruddy-ground Dove	144		90		18	149	112		50		563
Ruddy-tailed Flycatcher		4		2				9			15
Rufous Mourner		9		6				13			28
Rufous Piha		17		10				25		1	53

Species	HG	BR	VI	FE	FF	OP	ASP	WR	RT	TP	Total
Rufous-breasted Wren				3							3
Rufous-naped Wren	3		6								9
Rufous-tailed Hummingbird	38		18	4	14	6	23		23	4	130
Rufous-winged Woodpecker		1						2			3
Russet Antshrike		4		4				1			9
Scaled Pigeon	7				2		2		2		13
Scaly-breasted Hummingbird	2	4		4	2			7	1	2	22
Scaly-breasted Wren		7						6			13
Scaly-throated Leaftosser		4						1			5
Scarlet Macaw	30		13								43
Scarlet Tanager	2					2	3				7
Scarlet-rumped Cacique	7	35		30	13			20	3	19	127
Scarlet-thighed Dacnis				1	2						3
Scissor-tailed Flycatcher	1						1				2
Scrub Greenlet	2					2	1		1		6
Shining Honeycreeper	12		1	7	4				1	1	26
Short-billed Pigeon	12	54		70	35	14	17	63	18	13	296
Slate-colored Seedeater						1					1
Slate-headed Tody-Flycatcher	2				2			1			5
Slaty Antwren		1		1	2			3			7
Slaty Spinetail					1		4				5
Slaty-tailed Trogon	4	18		14	6		3	12	2	12	71
Smooth-billed Ani	19		28		3	25	39		33	2	149
Snowy Egret						3	11				14
Snowy-bellied Hummingbird				1							1

Appendix 1. (Continued)

Species	HG	BR	VI	FE	FF	OP	ASP	WR	RT	TP	Total
Social Flycatcher	69		74		18	2	36		37	5	241
Southern Beardless-Tyrannulet	7		1	1	5		2		8	2	26
Southern Rough-winged Swallow	3		3			24	26		2		58
Speckled Mourner								1			1
Spectacled Owl	2							2			4
Spot-crowned Euphonia	4	13		13	7			10		2	49
Spotted Woodcreeper		1		1							2
Squirrel Cuckoo				6			5	1			12
Streak-chested Antpitta		3						7			10
Streaked Flycatcher	25		22	10	11	24	21		12	31	156
Streak-headed Woodcreeper	10		6	1	6	6	11		7	8	55
Striped Cuckoo	3			1	2	5	4				15
Striped Owl						1					1
Striped Woodhaunter		1									1
Stripe-headed Brush-Finch		2									2
Stripe-headed Sparrow						2					2
Stripe-throated Hermit	5	8	2	17	2			6	3	2	45
Sulphur-bellied Flycatcher	3		2	1	1		2				9
Sulphur-rumped Flycatcher		9						7			16
Summer Tanager	10	1	1	2	6	1	4		4	4	33
Swainson's Thrush	5	4	1	6	7	1	2	3	5	2	36
Swallow-tailed Kite				2							2
Tawny-crowned Greenlet		33		12	1			39			85
Tawny-winged Woodcreeper		6		3	2			6		5	22

Species	HG	BR	VI	FE	FF	OP	ASP	WR	RT	TP	Total
Tenessee Warbler	17	6	13	9	55	13	23	4	13	25	178
Thick-billed Euphonia	6				5	3	3		4	2	23
Thick-billed Seed-Finch	1		1		6	19	27		4		58
Thrushlike Schiffornis		6		2				3			11
Tropical Gnatcatcher	5	3		16	15	6		9	1	12	67
Tropical Kingbird	119		117		36	49	73		86	12	492
Tropical Pewee	1		2		2		11				16
Turkey Vulture			16	5	1	9	3	2	1	5	42
Turquoise Cotinga	1										1
Variable Seedeater	125		85		56	180	177		88	8	719
Violaceous Trogon	4	2		10	6		1	2	1	4	30
Violet Sabrewing		1		2							3
Violet-crowned Woodnymph	1	13		10	5			10	1	6	46
Violet-headed Hummingbird	4	2		1		1		4			12
Wedge-billed Woodcreeper		38		24	3			22		9	96
Wedge-tailed Grass-Finch						2					2
Western Kingbird	12		20		10	29	15		19	1	106
Western Tanager				1							1
Western Wood-Pewee					1		3				4
White Hawk		1		2	1			1			5
White Ibis			2				15				17
White-breasted Wood-Wren								1			1

Appendix 1. (Continued)

Species	HG	BR	VI	FE	FF	OP	ASP	WR	RT	TP	Total
White-collared Seedeater	8		4			29	10		3		54
White-crested Coquette	2	1			2						5
White-crowned Parrot	87	28	42	48	30	28	71	5	25	48	412
White-necked Jacobin	15	3		9	10	3	5	1	7	8	61
White-necked Puffbird	3	1		5	1		3	3		1	17
White-ruffed Manakin		1			1						2
White-shouldered Tanager		12		1				12			25
White-throated Crake							2				2
White-throated Magpie-Jay	6					5				3	14
White-throated Shrike-Tanager		15		2				28			45
White-tipped Dove	50	2	15	19	88	40	42	5	49	34	344
White-tipped Sickbill		1									1
White-vented Euphonia		3		2				1			6
White-whiskered Puffbird				1				2			3
White-winged Becard				1	1		3			2	7
Wilson's Warbler					1	2		1		1	5
Yellow Tyrannulet	9		2		7	3	3		2		26
Yellow Warbler	29		32	1	9	36	20		8	17	152
Yellow-bellied Elaenia	22		10		5	2	23		23		85
Yellow-bellied Flycatcher		2			1					2	5

Species	HG	BR	VI	FE	FF	OP	ASP	WR	RT	TP	Total
Yellow-bellied Seedeater	3		1		1	3					8
Yellow-bellied Tyrannulet	2			2	1						5
Yellow-billed Cacique					9						9
Yellow-billed Cotinga							1				1
Yellow-breasted Chat					1						1
Yellow-crowned Euphonia	52		73	7	18	50	69	2	26	36	333
Yellow-crowned Tyrannulet	1			4			3				8
Yellow-faced Grassquit	1				2	3	11		9		26
Yellow-green Vireo	2				1						3
Yellow-headed Caracara	4		6		1	4	8			5	28
Yellow-olive Flycatcher	2			5		1			1	2	11
Yellow-tailed Oriole						1					1
Yellow-throated Euphonia	11		10	2	9	11	6		10	8	67
Yellow-throated Vireo	2		2	1	2	1	1				9
Yellow-throated Warbler					1						1
Total	*3373*	*1315*	*2879*	*1567*	*2094*	*2171*	*3095*	*1397*	*1892*	*1231*	*21015*

BR = Biological Reserves, FE = Forest Edges, FF = Forest Fallows, HG = Homegardens, OP = Oil palm, ASP = Agrosilvopastoral systems, WR = Wildlife Refuges, RT = Residential Tourism Projects, TP = Tree Plantations, VI = Villages.

REFERENCES

[1] A.G. Sanchez-Azofeifa, B. Rivard, J. Calvo, and I. Moorthy, *Mount. Res. Develop.* 22, 352 (2002).

[2] S. Polaski, E. Nelson, E. Lonsdorf, P. Fackler, and A. Starfield, *Ecol. Appl.* 15, 1387 (2005).

[3] K.S. Zimmerer, R.E. Galt, and M.V. Buck, *Ambio* 33, 520 (2004).

[4] L. Naughton-Treves, M.B. Holland, and K. Brandon, *An. Rev. Environ. Res.* 30, 219 (2005).

[5] N. Dudley, D. Baldock, R. Nasi, and S. Stolton, *Philo. Trans. Roy. Soc.* 360, 457 (2005).

[6] G. Powell, P. Wright, C. Guindon, U. Aleman, and R. Bjork, *Resultados y recomendaciones para la conservación de la lapa verde (Ara ambigua) en Costa Rica*, Centro Científico Tropical, San Jose, Costa Rica (1999).

[7] J.A. Hilty, W.Z. Lidicker, Jr., and A.M. Merenlender. *Corridor ecology: The science and practice of linking landscapes for biodiversity conservation,* Island Press, Washington (2006).

[8] V. Obando, *Biodiversidad en Costa Rica: estado del conocimiento y gestión*. INBio, Heredia, Costa Rica (2002).

[9] M.J. Groom, G.K. Meffe, and C.R. Carroll, *Principles of Conservation Biology,* Sinauer Associates, Inc. Massachusetts (2005).

[10] A.G. Sanchez-Azofeifa, G.C. Daily, A.S.P. Pfaff, and C. Busch, *Biol. Cons.* 109, 123 (2003).

[11] TNC and ASANA, *Evaluación ecológica rápida (EER): Corredor Biológico Paso de la Danta*. Puntarenas, Costa Rica (2000).

[12] O.E. Sala, F.S.C. III, Armesto, J.J. Berlow, E. Bloomfield, J. Dirzo, R. Huber-Sanwald, E. Huenneke., L.F. Jackson, R.B. Kinzig, A. Leemans, R.

Lodge, D.M. Mooney, H.A. Oesterheld, M. Poff, N.L. Sykes, M.T. Walker, B.H. Wlaker, and D.H. Wall, *Science* 287, 1770 (2000).

[13] A.F. Bennett, *Linkages in the Landscape: The role of corridors and connectivity in wildlife conservation.* Gland, Switzerland and Cambridge, UK (2003).

[14] D.A. Saunders, R.J. Hobbs, and C.R. Margules, *Biol. Cons.* 64, 185 (1991).

[15] A.B. Anderson and C.N. Jenkins, *Applying Nature's Design: Corridors as a strategy for biological conservation.* Columbia University Press, New York (2006).

[16] B.C. Klein, *Ecology* 70, 1715 (1989).

[17] W.F. Laurance, T.E. Loveloy, H.L. Vasconcelos, E. Bruna, R.K. Didham, P.C. Stouffer, C. Gascon, R.O. Bierregaard, S.G. Laurance, and E. Sampaio, *Cons. Biol.* 16, 605 (2002).

[18] G.V.N. Powell, J. Barborak, and M. Rodriguez, *Biol. Cons.* 93, 35 (2000)

[19] S.G.W. Laurance, *Ecol. Appl.* 14, 1344 (2004).

[20] E. Carrillo, http://*www.nacion.com* (2004).

[21] W.F. Laurance and R.O. Bierregaard, Jr., *Tropical forest remnants: ecology, management, and conservation of fragmented communities*. The University of Chicago Press, Chicago (1997).

[22] T.H. Ricketts, G.C. Daily, P.R. Ehrlich, and J.P. Fay, *Cons. Biol.* 15, 378 (2000).

[23] G.C. Daily, *Nature* 411, 245 (2001).

[24] J.B. Hughes, G.C. Daily, and P.R. Ehrlich, *Ecol. Letters* 5, 121 (2002).

[25] G.C. Daily, G. Ceballos, J. Pacheco, G. Suzan, and A. Sanchez-Azofeifa, *Cons. Bio.* 17, 1814 (2003).

[26] C. Gascon, T.E. Lovejoy, R.O. Bierregaard, J.R. Malcolm, P.C. Stouffer, H.L. Vascocelos, W.F. Laurance, B. Zimmerman, M. Tocher, and S. Borges, *Biol. Cons.* 91, 223 (1999).

[27] D.B. Lindenmayer and J.F. Franklin. *Conserving forest biodiversity: a comprehensive multiscaled approach*. Island Press, London, UK. (2002).

[28] H.R. Pulliam and B.J. Danielson. *Amer. Nat.* 137, S50 (1991)

[29] M.C. Horner-Devine, G.D. Daily, P.R. Ehrlich, and C.L. Boggs, *Cons. Biol.* 17,168 (2003).

[30] H.M. Pereira, G.C. Daily, and J. Roughgarden, *Ecol. Appl.* 14, 730 (2004).

[31] J.H. Lawton, D.E. Bignell, B. Bolton, G.F., Bloemers, P. Eggleton, P.M. Hammond, M. Hodda, R.D. Holt, T.B. Larsen, N.A. Mawdsleu, N.E. Stork, D.S. Srivastava, and D.A. Watt, *Nature* 391, 72 (1998).

[32] C.A. Lindell, W.H. Chomentwoski, and J.R. Zook, *Biod. Cons.* 13, 2419 (2004).

[33] R.F. Noss, *Cons. Bio.* 1, 159 (1987).

[34] R.F. Noss, In: *Landscape Linkages and Biodiversity,* W.E. Hudson, Island Press, Washington DC, pp. 27-39 (1991).

[35] R.T. Forman, *Land mosaics: the ecology of landscapes and regions.* Cambridge University Press, New York (1995).

[36] J.J. Tewksbury, D.J. Levey, N.M. Haddad, S. Sargent, J.L. Orrock, A. Weldon, B.J. Danielson, J. Brinkerhoff, E.L. Damschen, and P. Townsend, *PNAS* 99, 12923 (2002).

[37] G.R. Hess and R.A. Fisher, *Lands. Urban Plan.* 55, 195 (2001).

[38] M.E. Soule, In: *Landscape Linkages and Biodiversity*, W.E Hudson (Ed.), Island Press, Washington D.C. pp.91-104 (1991).

[39] K. Henein and G. Merriam, *Lands. Ecol.* 4, 157 (1990).

[40] P. Beier and R.F. Noss, *Cons. Bio.* 12, 1241 (1998).

[41] D.S. Simberloff and J. Cox, *Cons. Biol.* 1, 63 (1987).

[42] M.A. Boza, *Cons. Bio.* 7, 239 (1993).

[43] M. Wainwright, *The Natural History of Costa Rican mammals.* Zona Tropical Publication, Miami, USA. 381 p (2002).

[44] L.R. Holdridge, *Life Zone Ecology*, Tropical Science Center, Costa Rica (1967).

[45] C.D. Brockett and R.R. Gottfried. *Latin American Research Review* 37, 7 (2002).

[46] C. Kleinn, L. Corrales, and D. Morales. *Environ. Monit. Assess.* 73, 17 (2002).

[47] A.G. Sánchez-Azofeifa, C. Quesada-Mateo, P. González-Quesada, S. Dayanandan, and K.S. Bawa, *Cons. Bio.* 13, 407 (1999).

[48] A.S.P. Pfaff and G.A. Sanchez-Azofeifa, *Resource Energy Econ.* 26, 237 (2004).

[49] G. Hartshorn, *Costa Rica – Country environment profile: A field study.* Tropical Science Center. San Jose, Costa Rica, 123 p (1982).

[50] J.A. Tosi. *Los recursos forestales de Costa Rica*. Centro Científico Tropical. San Jose, Costa Rica. 17 p (1974).

[51] A. Wendland and K.S. Bawa. *J. Sustainable. For.* 3, 91 (1996).

[52] S. Sader and A. Joyce. *Biotropica* 20, 11 (1988).

[53] E. Oviedo, *www.nacion.com* (2006).

[54] A. Redondo, B. Vilchez, and R. Chazdon, *Revista Forestal Centroamericana* 36, 20 (2001).

[55] K. Welsh, *Water sustainability in the Baru and Guabo watersheds in Costa Rica: Assessing social and physical aspects within rural communities.*

Master's Thesis, School of Forestry and Environmental Studies, Yale University. New Haven, CT, USA. 99 p (2006).
[56] A. Redondo-Brenes and K. Welsh. *Tropical Resources* 24, 19 (2006).
[57] Q. Newcomer, *Endangered Sp. Update* 19, 186 (2002).
[58] A. Redondo-Brenes, *Tropical Resources* (Yale) 26, 7 (2007).
[59] O. Chinchilla-Mora and F. Mora-Chacón, In: *Memoria del taller-seminario espacies forestales nativas*, Universidad Nacional de Costa Rica, Heredia (2002).
[60] G. Stiles and A. Skutch. *A guide to the birds of Costa Rica*. INBio. Heredia, Costa Rica, (1989).
[61] R. Garrigues and R. Dean, *The birds of Costa Rica*. Cornell University Press, New York (2007).
[62] R.K. Colwell, *http://purl.oclc.org/estimates* (2005).
[63] J. Barlow, L.A.M, Mestre, T.A. Gardner, and C.A. Peres, *Biol. Cons.* 136, 212 (2007).
[64] O. Hammer, D.A.T. Harper, and P. D. Ryan, *Palaeontologia Electronica* 4, 9 (2001).
[65] N.S. Sodhi, L.H. Liow, and F.A. Bazzaz, *Ann. Rev. Ecol. Evol. Syst.* 35, 323 (2004).
[66] C.H. Sekercioglu, S.R. Loarie, F. Oviedo-Brenes, P.R. Ehrlich, and G.C. Daily, *Cons. Biol.* 21, 482 (2007).
[67] G.W. Luck and G.C. Daily, *Ecol. Appl.* 13, 235 (2003).
[68] M.A. Gray, S.L. Baldauf, P.J. Mayhew, and J.K. Hill, *Cons. Biol.* 21, 133 (2007).
[69] G.P. Garrity, *Agrofo. Syst.* 61, 5 (2004).
[70] F. Montagnini, *Environmental Services of Agroforestry Systems*. Hayworth Press, New York (2005).
[71] B.R. Scales and S.J. Marsden, *Environ. Cons.* 35, 160 (2008).
[72] I. Perfecto, R.A. Rice, and R. Greenberg, *BioSc*. 46,598 (1996) .
[73] R.A. Rice and R. Greenberg, *Ambio* 29, 167 (2000).
[74] E. Soini, *Afri. Zool.* 41,193 (2006).
[75] C.A. Harvey, and J.A. Gonzalez-Villalobos, *Biol. Cons.* 16, 2257 (2007).
[76] D. Tilman, K.G. Cassman, P.A. Matson, R. Naylor, and S. Polasky, *Nature* 418, 671 (2002).
[77] M.C. Amézquita, M. Ibrahim, T. Llanderal, P. Buurman, and E. Amézquita, *J. Sustainable. For.* 21, 31 (2005).
[78] F. Montagnini, In: Post-agricultural succession in the Neotropics, R. Myster, Springer, New York, pp. 265-295 (2008).

[79] D. Pezo and M. Ibrahim, *Sistemas silvopastoriles,* CATIE, Turrialba, Costa Rica,. Proyecto Agroforestal CATIE/GTZ, (1999).

[80] W. Yamamoto, I. Dewi, and M. Ibrahim, *Agric. Syst.* 94, 368 (2007).

[81] C.A. Harvey, C. Villanueva, J. Villacís, M. Chacón, D. Muñoz, M. Lòpez, M. Ibrahim, R. Gómez, R. Taylor, J. Martínez, A. Navas, J. Sáenz, D. Sánchez, A. Medina, S. Vílches, B. Hernández, A. Pérez, F. Ruíz, F. López, I. Lang, S. Kunth, and F. Sinclair, *Agricul. Ecosys. Environ.* 111, 200 (2005).

[82] C.A. Harvey, A. Medina, D. Sanchez, S. Vilchez, B. Hernandez, J.C. Saenz, J.M. Maes, F. Casanoves, and F. Sinclair. *Ecol. Appl.* 16, 19 (2006).

[83] C.A. Harvey, C.F. Guindon, W.A. Harber, D. Hamilton, and K.G. Murray, In: *Evaluación y conservación de biodiversidad en paisajes fragmentados de Mesoamérica*, C.A. Harvey and J.C. Sáenz (Eds). INBIO, Costa Rica, pp. 289-326. (2008).

[84] W. Francesconi, F. Montagnini, and M. Ibrahim, *J. Sustainable For.* In press (2010).

[85] M.G. Slocum and C.C. Horvitz, *Plant Ecol.* 149, 51 (2000).

[86] M.G. Slocum, *Ecology* 82, 2547 (2001).

[87] M.L. Enriquez, J. C. Sáenz, and M. Ibrahim, *Agroforestería en las Américas*, 45, 49 (2007).

[88] J.C. Sáenz, F. Villatoro, M. Ibrahim, D. Fajardo, and M. Pérez, *Agroforestería en las Américas* 45, 37 (2007).

[89] M. Ibrahim, F. Casasola, C. Villanueva, E. Murgueitio, E. Ramírez, J. Sáenz, and C. Sepúlveda, *J. Sustainable For.* In press (2010).

[90] G. Cardenas, C.A. Harvey, M. Ibrahim, and B. Finegan. *Agroforesteria en las Americas* 10, 1 (2003).

[91] C.A. Harvey and W.A. Habber, *Agrofor. Syst.* 44, 37 (1999).

[92] FAO, *Improving nutrition through home gardening.* FAO, Rome (2001).

[93] F. Montagnini, In: *Tropical Homegardens: A Time-Tested Example of Sustainable Agroforestry. Advances in Agroforestry 3,* B. M. Kumar and P. K. R. Nair, Springer Science, The Netherlands. pp. 61-84 (2006).

[94] M.E. Zaldivar, O.J. Rocha, E. Castro, and R. Barrantes, *Human Ecol.* 30, 301 (2002).

[95] N.W. Price, *The tropical mixed garden in Costa Rica*. Ph.D. Thesis. University of British Columbia, Canada (1989).

[96] R. Lok, In: *Huertos caseros tradicionales de América Central: características, beneficios e importancia, desde un enfoque multidisciplinario*. R. Lok (ed.), CATIE, Turrialba, Costa Rica, pp. 7-28 (1998).

[97] R. Lok, A. Wieman and D. Kass. In: *Huertos caseros tradicionales de América Central: características, beneficios e importancia, desde un enfoque multidisciplinario.* R. Lok (ed.), CATIE, Turrialba, Costa Rica, pp. 29-59 (1998).

[98] G. Budowski, In: *Tropical home gardens*. K. Landauer and M. Brazil (eds.), United Nations University, Tokyo, Japan, pp. 3-8 (1990).

[99] D.M. Griffith, *Cons. Biol.* 14, 325 (2000).

[100] J.F.Chace and J.J. Walsh, *Land. Urban Plan.* 74, 46 (2006).

[101] T. Tscharntke, C.H. Sekercioglu, T.V. Dietsch, N.S. Sodhi, P. Hoehn, and J.M.Tylianakis, *Ecology* 89, 944 (2008).

[102] J. Fischer, D.B. Lindenmayer, and A.D. Manning, *Frontiers Ecol. Environ.* 4, 80 (2006).

[103] R.L. Chazdon, *Persp. Plant Ecol. Evol. System.* 6, 51 (2003).

[104] D. Cusak and F. Montagnini, *For. Ecol. Manage.* 188, 1 (2004).

[105] F. Montagnini, *J. Sustainable For.* In press (2010).

[106] J.M. Carnus, J. Parrotta, E. Brockerhoff, M. Arbez, H. Jactel, A. Kremer, D. Lamb, K. O'Hara, and B. Walters, *J. Forestry* 1, 65 (2006).

[107] D. Lamb and D. Gilmour, *Rehabilitation and restoration of degraded forests*. IUCN, Gland, Switzerland and Cambridge, UK and WWF, Gland, Switzerland (2003).

[108] C. Orozco-Zamora and F. Montagnini, *Rest. Ecol.* 15, 453 (2007).

[109] R. F. Butler, F. Montagnini, and P. Arroyo, *For. Ecol. Manage.* 255, 2251 (2008).

[110] A. Redondo-Brenes and F. Montagnini, *For. Ecol. Manage.* 232, 168 **6w(2006).

[111] A. Redondo-Brenes, *New Forests* 34, 256 (2007).

[112] H.S. Lam Bent, F. Montagnini, and C.A. Finney, *J. Sustainable For.* In press (2010).

[113] F. Montagnini, A. Cerezo, H. S. Lam Bent, C. Finney, and T.J. Kim, *Yvyraretá (Argentina)* 15, 33 (2008).

[114] C. Vaughan, N. Nemeth, and L. Marineros, *Revista de Biologia Tropical* 54, 919 (2006).

[115] R.L. Chazdon, C.A. Harvey, O. Komar, D.M. Griffith, B.G. Ferguson, M. Martinez-Ramos, H. Morales, R. Nigh, L. Soto-Pinto, M. van Breugel, and S.M. Philpott, *Biotropica* 41, 142 (2009).

[116] E.C. Turner, J.L. Snaddon, T.M. Fayle, and W.A. Foster, *PLoS ONE* 3, 1572 (2008).

[117] L.P. Koh and D.S. Wilcove, *Nature* 448, 993 (2007).

INDEX

A

B

C

INDEX

A

B

C

S

T

U

V

W